Interoperable and Distributed Processing in GIS

ANDREJ VCKOVSKI
University of Zurich

UK	Taylor & Francis Ltd, 1 Gunpowder Square, London, EC4A 3DE
USA	Taylor & Francis Inc., 1900 Frost Road, Suite 101, Bristol, PA 19007

Copyright © Andrej Vckovski 1998

All rights reserved. No part of this publication may be reproduced, stored in a retrieval system, or transmitted in any form or by any means, electronic, electrostatic, magnetic tape, mechanical, photocopying, recording or otherwise, without the prior permission of the copyright owner.

British Library Cataloguing in Publication Data

A catalogue record for this book is available from the British Library.
ISBN 0-7484-0793-6 (cased)
ISBN 0-7484-0792-8 (paper)

Library of Congress Cataloguing-in-Publication Data are available

Cover design by Hybert Design and Type
Printed and bound by T.J. International Ltd, Padstow, Cornwall
Cover printed by Flexiprint, Lancing, West Sussex

WA 1176457 0

UNIVERSITY OF GLAMORGAN
LEARNING RESOURCES CENTRE

Pontypridd, Mid Glamorgan, CF37 1DL
Telephone: Pontypridd (01443) 482626

Books are to be returned on or before the last date below

- 7 APR 2000

2 2 MAR 2002

Contents

1	**Introduction**	**1**
	1.1 Problem statement	1
	1.2 Background	3
	1.3 Road map	5
I	**Interoperability**	**7**
2	**Interoperability in geoprocessing**	**9**
	2.1 Overview	9
	2.2 Data migration	10
	2.3 Integration and databases	14
	2.4 Data integration and geographical information systems	20
	2.4.1 Overview	20
	2.4.2 GIS and remote sensing	21
	2.4.3 Semantic heterogeneity	22
	2.4.4 Data quality	26
	2.4.5 Metadata	28
	2.4.6 System integration	30
	2.4.7 Environmental data management	31
	2.5 Review	33
3	**Virtual Data Set**	**35**
	3.1 Overview	35
	3.2 Interoperability impediments	36
	3.2.1 Syntactical diversity	36
	3.2.2 Semantic diversity of geographic information	36

		3.2.3	Diverse information communities	38
		3.2.4	Market forces	39
	3.3	Approaches for improvements		40
		3.3.1	Data definition languages	40
		3.3.2	Object-oriented models	41
		3.3.3	Semantic accuracy	45
	3.4	Open Geodata Interoperability Specification		46
		3.4.1	Overview	46
		3.4.2	Open Geodata Model	47
		3.4.3	Services Model	51
	3.5	The Virtual Dataset concept		54
	3.6	Review		60

II Essential models 63

4 Example: field representations 65
4.1	Overview	65
4.2	Continuous fields	65
4.3	Measurement of fields	67
4.4	Current representation techniques	74
	4.4.1 Cell grids	77
	4.4.2 Polyhedral tessellation	77
	4.4.3 Simplicial complexes	80
	4.4.4 Lattice or point grids	80
	4.4.5 Irregular points	81
	4.4.6 Contour models	82
4.5	Characteristics of the domain $\mathbb{D}$	84
4.6	Characteristics of the range	85
4.7	Analysis of continuous field data	86
4.8	Impediments	88
4.9	Review	89

5 Modeling uncertainties 91
5.1	Introduction	91
5.2	Uncertainties	94
	5.2.1 What is uncertainty?	94
	5.2.2 Basic concepts of modeling uncertainties	95
	5.2.3 Inference from uncertain propositions	97
5.3	Methods for the modeling of uncertainty	98
	5.3.1 Overview	98
	5.3.2 Probability theory	98
	5.3.3 Interval methods	100
	5.3.4 Fuzzy sets	103
	5.3.5 Dempster–Shafer theory of evidence	105

	5.4	Assessing the uncertainty of functional expressions	106
		5.4.1 Overview	106
		5.4.2 Gaussian error propagation	107
		5.4.3 Monte Carlo methods	109
		5.4.4 Other techniques	110
	5.5	Digital representation	111
	5.6	Numbers, units and dimensions	112
	5.7	Review	115

III Implementation 117

6 Case studies 119
  6.1 Introduction . . . 119
  6.2 POSTSCRIPT . . . 119
    6.2.1 Overview . . . 119
    6.2.2 Impediments . . . 121
    6.2.3 Document Structuring Conventions . . . 125
    6.2.4 Independence . . . 127
    6.2.5 Lessons learned . . . 127
  6.3 Internet and World-Wide Web (WWW) . . . 128
    6.3.1 Overview . . . 128
    6.3.2 Interprocess communication and network-level services . 129
    6.3.3 Scalability . . . 130
    6.3.4 World-Wide Web . . . 134
    6.3.5 Openness and standardization processes . . . 138
    6.3.6 The future: IPv6 . . . 138
    6.3.7 Lessons learned . . . 139
  6.4 Review . . . 140

7 Strategies 143
  7.1 Introduction . . . 143
  7.2 Component Object Model . . . 144
    7.2.1 Overview . . . 144
    7.2.2 Interfaces . . . 145
    7.2.3 Feasibility . . . 146
  7.3 Common Object Request Broker Architecture . . . 148
    7.3.1 Overview . . . 148
    7.3.2 Interfaces . . . 150
    7.3.3 Feasibility . . . 150
  7.4 Other approaches . . . 155
  7.5 Java . . . 157
    7.5.1 Overview . . . 157
    7.5.2 Design goals . . . 159
    7.5.3 Main features . . . 161

		7.5.4	Java as a distributed computing environment	170
	7.6	Review		180

8 Examples — 181
- 8.1 Introduction — 181
- 8.2 Uncertainty representation — 182
  - 8.2.1 Overview — 182
  - 8.2.2 A Java package for uncertain values: `vds.value` — 183
  - 8.2.3 Intervals — 186
  - 8.2.4 Probability distributions — 186
  - 8.2.5 Histograms — 188
  - 8.2.6 Dimensioned values — 192
- 8.3 Fields — 193
  - 8.3.1 Overview — 193
  - 8.3.2 A simple field — 195
  - 8.3.3 VDS and user interfaces — 195

9 Conclusion — 201
- 9.1 Retrospective view — 201
- 9.2 Prospective view — 205

A Java examples — 209
- A.1 Availability — 209
- A.2 Package contents — 210
  - A.2.1 Package `vds.value` — 210
  - A.2.2 Package `vds.geometry` — 211
  - A.2.3 Package `vds.field` — 211
  - A.2.4 Package `vds.test` — 211

References — 213

Index — 229

List of figures

2.1	Data exchange using a standardized format	12
2.2	Database integration	15
2.3	Five-level schema of a FDBS	17
2.4	Object- and field-models	24
3.1	OGIS diagram conventions	49
3.2	Essential model of OGIS feature type	50
3.3	Essential model of OGIS coverage type	52
3.4	Three-tier architecture of Virtual Data Set	59
4.1	Examples of selection functions with $\dim \mathbb{D} = 1$	71
4.2	Example of a selection function with $\dim \mathbb{D} = 2$	72
4.3	Discretization of continuous fields, s_i is a point	75
4.4	Discretization of continuous fields, s_i is a "pixel"	76
4.5	Cell grid with $\dim \mathbb{D} = 2$	78
4.6	Polyhedral tesselation with $\dim \mathbb{D} = 2$	79
4.7	Simplicial complex with $\dim \mathbb{D} = 2$	80
4.8	Point grid with $\dim \mathbb{D} = 2$	81
4.9	Irregular points, $\dim \mathbb{D} = 2$	82
4.10	Contour line model with $\dim \mathbb{D} = 2$	83
5.1	Union and intersection of two fuzzy sets	104
7.1	Interfaces in the component object model	145
7.2	Basic architecture of CORBA	149
7.3	IDL compiler, client stub and server skeleton	152
7.4	VDS implementation using HTTP	156
7.5	Java program compilation and execution	172

8.1	Value classes and interfaces	184
8.2	Example window of `HistogramGraphTest`, X Windows	190
8.3	Example window of `HistogramGraphTest`, Microsoft Windows 95	191
8.4	Field and geomtery classes and interfaces	194
8.5	Specifying a VDS to load	198
8.6	Example window of `GUIFieldTest`	198
8.7	Dataset with a configuration dialog	199

List of tables

2.1	Semantic heterogeneity: Relation `members`	19
2.2	Semantic heterogeneity: Relation `cantons`	19
4.1	Some examples of fields	68
5.1	Properties of different representation types for uncertain values	113
6.1	Berkeley sockets API	130
6.2	URL examples for various access schemes	136
6.3	Examples of MIME content types	137
7.1	Primitive data types in Java	161
7.2	Java packages	169

Acknowledgements

Part of this work has been funded by the Swiss National Science Foundation under contract 50-35036.92. The work has been accepted by the Philosophical Faculty II of the University of Zürich as an inaugural dissertation in the semester of Summer 1997 supported by referees Prof. Dr K. Brassel, Prof. Dr. R. Weibel and Prof. M. F. Goodchild.

I would like to express my gratitude to a number of people: Prof. Kurt E. Brassel for giving me a helpful and supportive working environment, Prof. Robert Weibel for many discussions, comments and useful suggestions at any time of the day, and Profs. Michael F. Goodchild and Michael F Worboys for their helpful comments and for acting as external reviewers.

I should also like to thank Frank Brazile for teaching me where to put a comma and for giving this book a grammatical face-lift, the staff of Netcetera AG who gave me the time to finish this thesis, even in very turbulent phases, Eva-Maria Lippert-Stephan and Felix Bucher, my colleagues in many projects (and night shifts), Heidi Preisig, who was a great support, and finally many others at the Spatial Data Handling Division and Department of Geography who supported this work in various ways.

Listings

3.1	Example of INTERLIS data definition	42
3.2	Example of EXPRESS data definition	43
6.1	"Hello, world" in POSTSCRIPT	121
6.2	POSTSCRIPT program with operator in procedure	122
6.3	POSTSCRIPT program with operator in comment	123
6.4	POSTSCRIPT program with conditional evaluation	123
6.5	POSTSCRIPT program randomly producing one or two pages	124
6.6	POSTSCRIPT Document Structuring Conventions	126
7.1	Example of CORBA interface definition language	151
7.2	Java base class `Object`	163
7.3	Assignment of primitive data types and object references	164
7.4	Interfaces in Java	165
7.5	Method main() of byte code in 7.6	173
7.6	Byte code generated from 7.5	173
7.7	Common interface `SomeInterface`	175
7.8	`SomeClass`, implements `SomeInterface`	175
7.9	Using `SomeClass` through interface `SomeInterface`	176
7.10	Socket server in Java	178
7.11	Socket client in Java	179
8.1	Output of `IntervalTest`	187
8.2	Output of `UniformTest`	189
8.3	Output of `HistogramTest`	189
8.4	Output of `DimensionTest`	193
8.5	Output of `FieldTest`	196

Foreword

Language is our primary means of communication, yet humanity is divided into between three and four thousand speech communities, and in most cases it is impossible for members of two different speech communities to understand each other. There are problems of communication even within speech communities (as in Churchill's famous description of the U.S. and the U.K. as "two nations divided by a common language"). Pictures present no such problems, and a picture of a lone protester in Tienanmen Square is instantly meaningful to everyone, whatever the language of the commentator. A picture of Earth from space similarly transcends language, speaking directly to the humanity in all of us.

Maps fall somewhere between these two extremes. Almost any sighted person can draw useful information from a street map of London, though the names of features will be inaccessible to someone not familiar with the rudiments of English pronunciation and the Roman alphabet. Similarly, a foreigner in Tokyo might have no problem with understanding the city's layout from a map, but might be totally unable to recognize place names in Japanese characters. Common cartographic convention makes these two maps partially interoperable, though there may be no interoperability of language. This set of common cartographic conventions has developed over the past two centuries, and it has allowed maps to be exchanged freely as a geographical *lingua franca*. There has been no need (at least in recent years) for the British Ordnance Survey to make maps of the U.S. because U.S. maps are available and fully informative to British users. The number of map communities, in the sense of communities sharing a common geographic language, has been far less than the number of speech communities.

Geographic information systems (GIS) provide us with a way of capturing geographic information in digital form, and manipulating, sharing, and displaying it in myriad ways. The contents of a map can be flashed around the world in digital form at the speed of light, vastly increasing the usefulness and value of our geographic knowledge of the world. So one might think that the advent of GIS would have increased the homogeneity of cartographic convention, just as increased travel has led to the dominance of English as the international language of tourism and commerce. Surprisingly, exactly the opposite seems to have happened. There turned out to be many different ways to convert the contents of a map to digital form, and many ways emerged to specify the various operations one can perform using a GIS. The command languages that were developed to control GIS were consequently almost completely incompatible. Apparently simple terms like "layer" and "object" turned out to mean different things in the context of each GIS. Moreover, commercial interest seemed to argue for uniqueness, as each software company tried to create its own market niche and to prevent its users from making an easy switch to a competitor's products. The result was a proliferation of GIS communities, each defined by one set of products and unable to communicate easily with others.

But times have changed, and today's users of computers are no longer willing to endure the long process of training needed to master the idiosyncrasies of one vendor's design, when the task that needs to be accomplished seems conceptually simple. Vendors have realized that openness, in the sense of easy compatibility, is a strategy whose advantages in the marketplace can outweigh those of the opposite strategy of secrecy and entrapment. The GIS industry has begun a long process of convergence towards open specifications and standards that will eventually allow one system to talk to another, and one vendor's components to be replaced by those of another vendor. These practices have dominated the electrical and automobile industry for decades, and the GIS software industry is finally beginning to see their advantages.

Andrej Včkovski is one of the foremost researchers in this new arena of interoperability and open GIS. He was among the first to recognize the true implications of interoperability, in the form of a new kind of GIS that would be easier to learn about and use. A common language between GIS is necessarily a simple language that replaces the arcane complexities of each system's proprietary approaches with one that is solidly grounded in theory. By removing all of the unnecessary details of technical implementation from the user's view, it allows interaction with a simple conceptualization. This book breaks new ground in showing how this basic principle works, using well-chosen examples. As a piece of research it is still some distance ahead of the industry, so in that sense this book offers a view of the GIS that is yet to come. At the same time the approach is very practical, and it is easy to see how it can be implemented, given the clarity of the author's presentation and the simplicity of the ideas. The publication of this book should help to move that process of implementation along faster.

Interoperability should make GIS easier to use and easier to learn about. It should make it possible to transfer skills acquired on one system to another, and make it easy to transfer data created for one system so that it can be manipulated by another. All of this will have exciting impacts on how we learn about GIS, and how we make use of it. Thus far all we know is that future technology will make it possible to store data in one location, process it in another, and control the process from a third. It will also make it possible to unbundle GIS into simpler components that can be reassembled at will. But we have very little idea of how to take advantage of that potential, or of what it implies for practices of use, archiving, and analysis. In a world in which computing can occur anywhere, in any sequence, decisions will still have to be made about where to compute, and in what order. Entirely new choices become possible, and there will have to be rational grounds for making them, and for the new arrangements that will have to be put in place. As so often in the past history of GIS, new capabilities create new potential and new uncertainty. In that sense this book is a stimulus to a new round of research that will address the deeper implications of achieving interoperability, and a stimulus to educators to think about how to prepare the next generation of GIS users to work in a very different world.

Michael F. Goodchild
University of California, Santa Barbara

CHAPTER ONE

Introduction

1.1 Problem statement

This thesis is about integration. The term *integration* stems from the Latin word *integrare* which is etymologically based on *in-tangere*, a negation of *tangere*. Therefore, integration has to do with *untouchable* things and means – according to the definition of the Oxford Dictionary – the *bringing together* of *parts*. Actually, there are two meanings which differ slightly: on the one hand, integration means putting together parts *of a whole*, on the other hand it can mean bringing together *independent parts*. The main difference lies in the *autonomy* of the parts, i.e., are the parts *intended* to be brought together and only meaningful as a whole, or do they have a meaning by themselves?

This thesis is about *data* integration. The parts which have to be integrated are data, and – more specifically – *spatial* data. Spatial data means that the data contain elements which can be interpreted as locations of a vector space. The space could be any abstract space. However, only concrete, physical spaces which are mostly subspaces of four-dimensional space–time are considered here. The spaces of interest are related to the *earth* and those that meet the requirements of earth sciences in terms of position, connectivity and scale, e.g., spaces representing the earth's surface.

The references to positions in such spaces makes spatial data special with respect to the integration methods and impediments. Some characteristics of the spaces considered both *create* integration problems and *provide solutions* to these problems. Such a fundamental characteristic is a notion of *continuity*, that is, the spaces considered here are *topological spaces* or *manifolds*. Continuity means that there are concepts of connectivity and, in most cases, there is a metric available. This means in turn that two elements of such a space share common properties and that functions can be defined, for instance, to denote the *distance* between two elements. Moreover, intuition indicates that those properties have more in common if their distance is small.

The integration of data, as will be seen later on, needs the identification of common entities within the parts. Spatial references as common entities are a natural choice also because physical evidence says that a spatial location usually is only occupied by one object. Two objects having the same spatial reference ("same place same time") are likely to be related to each other. So, the impediments mentioned before have to do with determining if two spatial references are "more or less the same" even if they are not exactly the same. Likewise, integration benefits accrue from the possibility for estimation within the spatial neighborhood of elements given in the data. This allows for a *meaningful estimation* of additional data elements, referring, for instance, to nearby spatial locations. Such data will be called *virtual data* later on and form a core concept of this thesis.

Integration, in general, makes sense only if the *integrated* thing is more than the sum of its parts, that is, integration implies some sort of interaction between the parts. The parts have in most cases a certain autonomy, and interaction is not possible *a priori*. Interaction needs to be enabled by adherence of the parts to some kind of common rules. If states are integrated to form, for instance, a federation, there are rules needed to drive the federation. For example, a common constitution could define some rules for exchange between the states, such as a common currency to support trade from one state to another. If the native language within all states were different, a common language could be created for the exchange of information. Each state would have to employ only translators from their native language into the common language and vice versa, instead of translators from and to every other language of the federation. Such a *lingua franca* would be the basis for any information exchange between the states in the federation. The requirements for the *lingua franca* in place are very high. It needs to be at least as *expressive* as the local languages in every state so that every statement in every language can be translated into the common language. It also needs to be *flexible* and *extensible* in order to cope with new developments in the local languages. Next, it needs to be *well-defined*, that is, the semantics of the language must be clear and accurately defined to avoid misunderstandings and unnecessary legal disputes. And finally, it needs to be *simple*, so that many people can learn and use it and many translators can be trained, as opposed to a few highly-paid specialists.

These requirements are contradictory to some degree. There is a trade-off between *expressivity* and *simplicity* and also between *accuracy* and *flexibility*. However, there is a chance that these objectives can be met if the language is based on a clear foundation, avoiding all exceptions and ambiguities of the "legacy" languages. In computing systems, such a process is sometimes called *re-engineering* and means a reconsideration of the problems to be solved and a reconstruction of a system based on new insights.

This thesis basically proposes two things. It argues for the necessity to have, first, a re-engineered *lingua franca* for the integration and exchange of spatial data, and second, *intelligent* and *reliable* translators. What does intelligent and reliable mean in that context? Intelligent will say that a translator can think and also translate things which are not explicitly represented within a message but which can be easily inferred from that message. For example, consider the message "there are no animals allowed which are bigger than a dog". A intelligent translator could infer that elephants are not allowed, even if it is not explicitly stated in the message. Such a translator is much more useful because he or she could also be helpful in situations where a strict one-to-one translator fails. "Reliable" means that a translator produces messages one can count on. This also means that the translator must be able to communicate vagueness and uncertainty which might be present in the message being translated. The translator in this thesis is the central part of a virtual data set. A virtual data set can be seen as a message with its own translator. Every message is associated with its own translator. The translator is an expert on the message and can produce a variety of reliable statements in the *lingua franca* based on the understanding of the message. The important thing is that these statements are not created once and written down, but are inferred when the translator is asked, that is, upon request. It allows the translator to answer many more questions than if he or she were to have written down some of the most frequently asked questions about the message. For example, the translator could also infer that "mammoths are not allowed" even if this is not a frequently asked question. There might be many different ways a translator produces such *virtual* statements. For example, one way might be based on knowledge about the message which is carried around with the translator. Or, the translator might call a friend who is an expert for some questions. The important thing is that the translator's client *does not need to worry about* how the statement is obtained. The client just knows *how* to ask the translator a question. That is, there is a clear separation of responsibilities between the information consumer and information producer.

In summary, this study pursues the following goals:

- analysis of the principles of interoperability and data integration in the context of geoprocessing, and

- discussion of strategies and techniques to implement interoperability schemes.

1.2 Background

This thesis started within a project which was part of a larger group of projects dealing with various topics in the context of global change research, and more specifically, with the climate and environment in alpine regions. The setting consisted of several groups from different disciplines such as: physics, biology, geography, sociology and economy; the multi-disciplinarity of the project was one of its special characteristics. The subproject dealt with methodological questions regarding the migration of data between the various research groups and the integration of data produced there. The main objective was to support new scientific insights and findings by a technological framework that eases data exchange and integration and that promotes scientific collaboration across disciplines.

The analyses of the scientists' requirements and their typical problems with data exchange and integration showed that there were various general problems known from multi-disciplinary work which impeded data migration and integration such as different nomenclature and so on. Information exchange on the basis of research articles and other text material worked quite well. After all, all groups were doing research on various aspects of the same problem. Data migration and integration, however, could not be established in the same manner at all. The problems were sometimes regarded as being of a purely technical nature and a necessary evil. Further analysis showed that there are more fundamental issues within the data integration problem than mere data format conversions. Thus, the focus of the project changed towards generic studies of scientific data management and procedures for scientific data integration. The topic of the overall project motivated a *pars pro toto* approach considering data from environmental sciences as a case study for more general spatial data integration. Environmental data are actually a very good example for many spatial data integration problems because data sets represent *four-dimensional phenomena* (making them a general case), data are *uncertain* (data quality issues are important) and there is a strong *need* for data integration. The need is due to various reasons. First, many environmental problems can only be addressed in an integrated, multi-disciplinary way. Second, some physical quantities needed for a specific simulation might be derived from other measurements which are easier to sample and already available in digital form. Third, environmental sciences often work with mathematical simulation models of environmental processes. These simulations need data for model identification, parameterization and validation.

The technical background of this thesis is in the field of *Geographical Information Systems (GIS)*, information systems serving as a sort of headquarters in federations of spatially referenced data. GIS has the potential to provide a basis for environmental research and to be a useful tool for data management and analysis. Typical GIS software commercially available at the time of writing usually does not provide the requirements for a *lingua franca* as discussed above. The current software is large, inflexible, complicated, oversized and specialized. This is due to general trends in the computing industry during the past decades to provide proprietary and monolithic solutions because there was no other way of meeting the user's requirements. Hardware and other limitations made it impossible to refrain from using proprietary technologies. Also, using proprietary technology was meant to produce a secure market and bind clients to one or other vendor's products. Experience over the years has shown, however, that *openness* can be the basis of commercial success as well and most software vendors did change their strategy towards small, interoperable, component-based systems. Customers benefit in various ways from such products. Software components can be assembled in a dedicated way to exactly meet the user's requirements. That is, openness allows specialization. The competition leads to better soft- and hardware quality and lower prices. And finally, it allows for creative and innovative solutions which can be embedded into existing systems.

1.3 Road map

This thesis addresses integration and interoperability issues with a "longside", incompletely treating issues on various levels of abstraction. Chapter 2 gives an overview of data migration and integration. It discusses these issues in the context of GIS as well as database management systems. Semantic heterogeneities are presented as a major type of impediment which has to be overcome with a successful integration strategy.

Chapter 3 analyzes the impediments and presents two components of an improved data integration strategy. One is the *Open Geodata Interoperability Specification (OGIS)* which is an attempt to provide a *lingua franca* for the interoperability of spatial data. The other component is the translator-within-the-message principle called *Virtual Data Set (VDS)*. One of the important strategies to overcome semantic heterogeneities is a better understanding of the data semantics, i.e., the *model* defining the relation between abstract entities such as numbers in a data set and "reality". In some object-oriented design techniques this is called an *essential model*, as discussed in chapter 3. Actually, the building of essential models is *the essential* scientific objective: "The purpose of an essential model is to *understand* a situation, real or imaginary" (Cook & Daniels, 1994, p. 12, emphasis added). In natural sciences, parts of the essential model are often *mathematical* models, providing a mathematical abstraction of the situation under consideration. These models can serve as a basis for a subsequent implementation on computers. Chapters 4 and 5 show two example of such essential models. These two examples are of paramount importance for scientific data management.

The first example in chapter 4 discusses *continuous fields* and provides an *essential model* for fields. Continuous fields refer to physical quantities which are related to spatial references from a manifold, that is, they are functions of the elements of that manifold. The second example, in chapter 5, analyzes the thematic components of spatial data sets and discusses *uncertainty* issues when managing numerical values representing physical quantities. This leads to an essential model for the digital representation of numerical values which are affected by uncertainties (i.e., *all* measurements).

The implementation of the concepts introduced in chapter 2 and exemplified in the following chapters offers various alternatives. As an introduction to the discussion of implementation alternatives, two case studies are presented first. These case studies in chapter 6 are POSTSCRIPT and computer networks based on the Internet protocols. The rationale for the case studies is to show aspects of interoperability found within two computing systems having a *lingua franca* on different levels. POSTSCRIPT is a programming language which is used to communicate graphical content ("page description") between data producers and data consumers. POSTSCRIPT has a high expressivity compared with other graphics exchange mechanisms and was designed with explicit interoperability goals in mind. The case study shows the potential and also the danger of a highly expressive environment. The second case study has its focus on the Internet. The Internet has grown to a size which starts to be comparable to traditional telephone networks[1], and it works. It is a good example of various *successful* integration and interoperability aspects.

Based on these case studies, a few implementation alternatives are discussed in chapter 7. These include the Object Management Group's *Common Object Request Broker Architecture (CORBA)*, Microsoft's *Component Object Model (OLE/COM)* and Java. These three alternatives are *not exclusive* and have slightly different coverage. Java, for example, is not only a software system for distributed object environments but also a programming language with extensive class libraries. Actually, using Java as a distributed object environment will typically also include CORBA or another mechanism for the communication between object user and object implementor. This chapter shows that Java has a number of properties which make it a favorable choice for the implementation of Virtual Data Sets.

Chapter 8 combines the essential model developed in chapters 4 and 5 and the implementation strategy derived in chapters 6 and 7. Several aspects of an implementation based on Java are sketched with some running example programs in order to show the feasibility of the approach.

The conclusion in chapter 9 summarizes the wide range of topics covered in this thesis and proposes a set of issues which need to be addressed in future research.

As indicated before, the problem area is not fully explored. It rather tries to "scratch the surface" of the problem area to a deliberate depth. One important objective is to achieve some understanding of the underlying principles. Being broad sometimes helps to avoid omissions of important topics. Being deep is sometimes necessary to "touch the ground", to show that a proposed solution is possible and not just many words about nothing.

It is good practice for scientific research to raise questions. It is, however, the ambition of this thesis to give some significant answers as well.

---

[1] Have you ever wondered about the high reliability of telephone networks? These networks are very, very large but rarely fail.

I
Interoperability

CHAPTER TWO

Interoperability in geoprocessing

2.1 Overview

Interoperability issues have become a major concern within computing environments in the last decade. The use of computers has grown beyond specialized, isolated applications and produced a need to exchange information and other resources between computer systems – and eventually – computer users. *Interoperability* refers in general to the ability of various autonomous systems to bring together parts and to operate in collaboration. In most cases this means the *exchange* of meaningful *information*. Within computing environments there are many different aspects of interoperability, such as:

- Independent applications A and B running on the *same* machine and operating system, i.e., interoperability through a common hardware interface.

- An application A reading data written by another application B, i.e., interoperability through a common data format.

- Application A communicating with application B by means of interprocess communication or network infrastructure, i.e., interoperability through a common communication protocol.

Besides technical issues, there are also interoperability topics at much higher levels of abstraction. Consider, for example, two autonomous libraries using the same hardware and software to manage their directories of books. Both libraries agree to exchange their data in order to provide information about books available in the other library. The computer system used is the same, so there are no technical barriers for the migration of the data. However, there are many semantic barriers to overcome. For example, it might be that the number spaces used for call numbers by both libraries are not disjoint; or, the register entries of the libraries are different. This means that there are impediments other than the technical ones, and often, these impediments are the more difficult to overcome. In the case of the libraries, both institutions would probably work out a contract to define, for example, the number spaces to be used by each library and a common set of register entries.

As can be seen from both the technical and non-technical examples, the key issue in achieving interoperability or compatibility[1] is a *contract*[2] between interoperating parties defining a set of agreed common features.

Often, there are many parties involved within such an interoperable environment. For example, many computers might be connected to the same physical network and therefore interoperate in the way they access the cables, e.g., signal levels and so on. Or, many libraries might agree on mutual data exchange, i.e., constituting a large virtual library. Often, not all members of the interoperable environment are known at the time the environment is set up. Additional computers might be connected to the network later, or other libraries might join the "virtual" library at a later point in time. In such situations it is obviously not sensible to base interoperability on mutual contracts between two parties. With n parties this would lead to $n(n-1)$ signatures (every party would sign $n-1$ contracts) and for every new party joining $2n$ new signatures are needed. The approach usually taken is to agree on a single common contract which is signed by all parties. This needs n signatures for n parties, and each new member just needs to provide one signature. In the context of computer systems we call the common contract a *standard* which defines, for instance, a data format, an application programming interface (API), or a communication protocol. The signatures symbolize a party's *compliance* with the standard.

This chapter discusses interoperability issues as they arise in spatial data handling, in general, and environmental data management in particular. The next section discusses the importance of *data migration* as a low-level interoperability issue. This is followed by sections covering *data integration* and the particularities of these issues in the context of *environmental data management*.

2.2 Data migration

The growing use of *Geographical Information Systems (GIS)* is accompanied by intense efforts for standardization of structures and methods for the interchange of spatial data (Moellering, 1991). These standardizations lead to

---

[1] *Interoperability* and *compatibility* will be used interchangeably in this thesis. *Interoperability* probably better fits the overall discussion because it exhibits a certain *dynamic* character.

[2] Using the word *contract* provokes an intended similarity to an often used design principle for object-oriented systems which is called *design by contract* (Meyer, 1992).

well-defined data formats which are used as a *common contract* as discussed above. In the context of GIS, data exchange is particularly important because it is typically expensive to create new datasets[3]. In order to migrate data using a standardized data format between two systems both systems have to provide corresponding translators, i.e., their *signature* to the contract (see figure 2.1). As mentioned above, the use of a common "neutral" format reduces the number of necessary translators from $n(n-1)$ to $2n$ (Lauer *et al.*, 1991; Oswald, 1993).

A data transfer standard defines – in the end – what the meaning of a sequence of binary numbers means, and, given a "meaningful data representation", how this representation can be transformed into a sequence of binary numbers. Of course, most spatial data transfer standards do not cover representational details down to the binary level. Instead, a collection of additional standards is adopted which regulates details not covered by the standardization itself. For example, it is common that the encoding of *text* is not defined in detail within such a spatial data transfer standard. The text encoding is only given by, for example, stating that "it has to be in ASCII".

As with the interoperability examples above, spatial data transfer standards are established on various levels of abstraction. Some standards might, for example, only define geometrical primitives and leave their semantics open. In such a "geometry-only" standard, there is no difference between a river and a road in terms of the exchange format[4]. Other formats include detailed semantic catalogs of types of features and attributes allowed (DIGEST, 1991).

Standards development often lags some way behind the growth of a technology (Cassettari, 1993, p. 121). This often leads to situations where data formats used by the most successful vendor at a specific time become *de-facto* standards. For various reasons, however, it is unacceptable for international and national organizations, such as the *International Standards Organization (ISO)*, to adopt such standards. For example, data formats provided by vendors often are trade-marked or subject to other restrictions based on the vendor's intellectual property. Besides legal issues there are also other factors which motivated national organizations to develop their own standardization. The management of spatial data developed national particularities, especially in cadastral and defense contexts. With the movement from the analog to the digital domain many of these particularities were maintained in digital spatial

---

[3] However, existing datasets are sometimes expensive, too. It has been shown that the price of datasets does influence the demand much like other products (Lauer *et al.*, 1991).

[4] The river-object and the road-object might be labeled differently or have different sets of attributes, so that a human observer could determine whether it is a river or a road. The important thing is that this difference is not defined on the data format level. It is impossible to write an automatic "river-detector" using only the information provided in the format and no additional knowledge such as "if it has a depth, then it is probably a river; if the number of lanes is specified then it is a road". If additional information about the topology would be available then one could test for cycles in the graph of the segments. Rivers typically are represented in a directed and acyclic graph unless there are channels and the like.

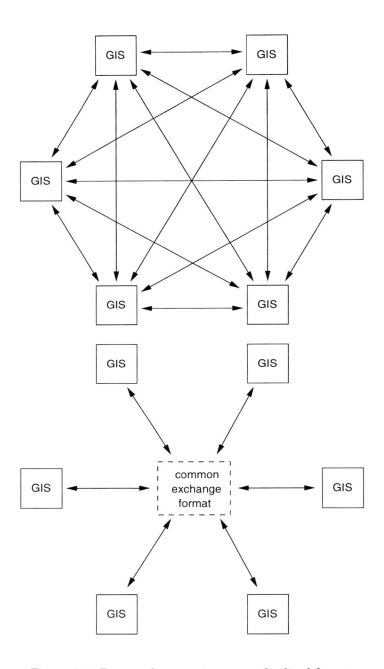

Figure 2.1: Data exchange using a standardized format

data management, and consequently, are important for data interchange. It might be, for example, that data interchange in Switzerland requires multi-lingual text representation but it is not at all important to have a sensible differentiation between various types of lagoon. Other countries might not be confronted with multi-lingual issues, but might rely on a good characterization of coastal zones. The variety of requirements lead to many different standards currently available (Cassetari, 1993, p. 142; Sorokin & Merzlyakova, 1996). Current developments within ISO, however, try to come up with a standardization which is acceptable for many spatial data providers and users, as well as system vendors. Instead of having separate standards for the exchange of letters, articles, memoranda, English books, German books, and so on, a standardization for generic "text" is approached. Spatial data transfer standards typically address the following objectives:

- The data model used must be *expressive* enough to allow loss-free translation from and to a majority of applications.

- The format should be *platform-independent*, e.g., not rely on a specific byte-ordering or text encoding.

- The format must allow *complete* data description, i.e., it should contain all information required to interpret it.

- The implementation of import- and export-translators should be as *easy* as possible in order to promote vendor acceptance.

- A standardization needs to be *extensible* to assimilate new developments.

- Datasets of different sizes need to be supported, i.e., the standardization must be *scalable*.

- Broad acceptance by many users and institutions is needed to achieve a substantial usage.

Some of these requirements conflict. For example, there is a clear trade-off between expressiveness and ease of implementation. The more flexible a format the higher is its complexity and, therefore, its implementation cost. It has to be noted, however, that there is an asymmetry between translation *into* and *from* a standard data format. In most cases, it is much simpler to translate an internal representation into a standard format than to read a standard format and then convert it to an internal representation. The latter needs a full understanding of the data format while the first needs only the subset necessary to represent the particular data types represented in the internal system. The POSTSCRIPT language discussed as a case study in chapter 6 is an example of a standard format with high expressiveness. It is fairly straightforward for applications to create POSTSCRIPT code whereas interpreting, i.e., importing, POSTSCRIPT code is a complex task.

Standard formats are sometimes also used not only for the migration of data but also as a method to enable persistence on secondary storage. Within remote sensing and other image processing applications, it is even *common* for datasets to be kept in a standard format within a file system. Translation from the format into an application-specific representation happens when the datasets are loaded and processed. However, standard data formats are designed in most cases for data migration. This means that the formats need to be able to use storage systems with a limited functionality such as magnetic tapes. That is, the formats do not require more sophisticated storage managers such as a file system, or even a database management system (DBMS), to be in place. Using standard data formats as a data repository does not make use of features offered by, for instance, a DBMS. It has been shown that such standardized data formats can be too inflexible as a repository for large spatial databases (Shaw *et al.*, 1996).

2.3 Integration and databases

In the field of database theory, *data* or *database integration* is an often discussed interoperability topic. There is a growing need to access data simultaneously from various distinct databases, so-called *component databases (DBS)*. These are, for example, databases within various divisions of large organizations which were developed individually but have to be accessed now on an integrated level. An approach often taken is to re-engineer the databases, create a new, single database and *migrate* all data into the new database. This approach yields the highest degree of integration, eliminating any heterogeneities by setting up a new, homogeneous system. In many cases, however, it is not possible or desirable to migrate all data into a homogeneous system owing to various reasons, such as:

- Investments in existing applications need to be preserved; for example, legacy applications depending on component databases.

- It might not be desirable to relegate authority for particular parts of the data to a central system.

- Size and performance considerations might be prohibitive for a migration approach.

There are several levels of integration possible if a migration approach is not feasible. Sheth and Larson (1990) provide a taxonomy of such *multidatabase systems (MDBS)*. Major differentiations relate to *homogeneity* and *autonomy*. A MDBS is *homogeneous* if the component DBS are managed by the same type of software and *heterogeneous* otherwise. A MDBS is a *federated database system (FDBS)* if the autonomy of the component DBS is preserved.

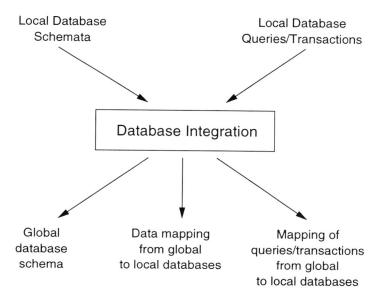

Figure 2.2: Database integration (Batini *et al.*, 1986)

The case of autonomous component systems in a federation (FDBS) fits in the context of this thesis. There are various methodologies for database integration of component DBS into a FDBS (Batini *et al.*, 1986). The goal of this process is basically to find mappings between the schemata of the component systems and a federated schema and to define mappings between the corresponding local and global operations (see figure 2.2).

Sheth and Larson (1990) propose a five-level schema to describe the architecture of a FDBS. Translators provide *schema translation* and *schema integration* between the different schemata. The five schemata identified are:

Local schema Conceptual schema of a component DBS, expressed in the native data model of the component DBS.

Component schema Translation of the local schema into the *canonical data model* of the federation, i.e., the model used for the federated schema.

Export schema A subset of the component schema which contains all data being exported to the federation, i.e., omitting "private" data.

Federated schema Integration of multiple export schemata (sometimes also called *global schema*).

External schema The external schema is used for applications and users of the FDBS. This schema might have additional constraints and access restrictions.

If a particular component DBS uses the same data model which is used as the canonical data model, then it is not necessary to differentiate between the local schema and the component schema. Also, if the entire component DBS is exported, i.e., there are no private data, then the export schema is equal to the component schema. Figure 2.3 shows these schemata. Translation processes convert between them. In a similar manner, operations are translated from the FDBS to the component DBS, i.e., a translation of commands and queries to the FDBS into local operations on the component DBS.

The federated schema – expressed in the canonical data model – is a core component of this architecture and represents the common contract in the federation[5]. The schema translation and schema integration processes (Johannesson, 1993) are each component DBS's signature to the contract, i.e., by being able to translate to and integrate the local schema into the federated schema a component DBS takes part of the federation. The integration process is based on the *identification of common entities* within two component DBS. This identification is possibly both on the intension and the extension of the local schemata, i.e., intensional parts of the schema (attributes) can be matched as well as parts of a schema's extension. The integration into a federated schema tries in most cases to produce a *minimal, complete* and *understandable* global schema (Batini *et al.*, 1986, p. 337), complete (and correct) in the sense that the integrated schema must contain all concepts from the local schemata correctly. *Minimality* means that concepts being present in several local schemata are represented in the integrated schema only once. *Understandability* refers to the optimal choice between various possible integrated (federated) schemata, i.e., the optimal choice is a schema which can be easily understood. The integrated or federated schema often contains *generalizations* of the concepts found in the local schemata. Consider for example two relations, *employee* and *customer*, in two separate local DBS. The integration might lead to a relation, *person*, in the federated schema which contains the matching intension of *employee* and *customer*, e.g., name, age, phone number and so on. The concept of generalization is important in the context of spatial information systems as will be shown below. It basically means that integration can be achieved or more easily achieved if concepts on a higher abstraction level are used for the identification. In practical applications there are, however, many problems induced by the *autonomy* and the *heterogeneity* of the component DBS. Depending on the level of autonomy, various problems can occur, such as:

- The local database might violate constraints existing in the federated schema.
- The local schema might be changed.
- Local data might be changed without notification of other component DBS, possibly influenced by new data.

---

[5] However, there are approaches without a federated schema (Litwin *et al.*, 1990).

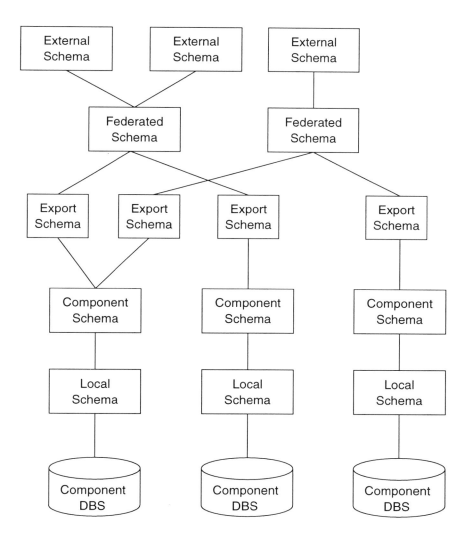

Figure 2.3: Five-level schema of a FDBS (Sheth & Larson, 1990)

Such problems are fundamental and cannot be overcome without giving up part of each component's autonomy. This means that with full autonomy, it is inevitable that there are *inconsistencies* in the federated system. It is important to note that such inconsistencies often cannot be resolved in the context of FDBS, unlike the case of data migration. When migrating data from component DBS into a centralized system, it is also necessary to perform the tasks of schema integration and translation and it is also common that inconsistencies may result since the component systems were autonomous before migrating the data. The important difference is, however, that the process of resolution of inconsistencies and the homogenization happens *once*, i.e., in a controlled way. Depending on the data volumes, it might be feasible, for example, to adopt manual interactive inconsistency resolution procedures. In FDBS, however, it is necessary to cope with inconsistencies. It will at times be impossible to decide, for example, which of a set of conflicting alternatives is true. As will be shown later on, this is particularly important in the context of spatial data. Not only may inconsistencies exist between components of an integrated system, but there may also be irresolvable inconsistencies within a single component DBS.

The heterogeneities between the component DBS lead to problems which are common both in data migration and integration. These heterogeneities exist on various levels, such as:

- Different computer hardware and operating systems.

- Different DBS management software.

- Different data models used, e.g., relational models, networks models, object-oriented models.

- Semantic heterogeneities.

The first three types of heterogeneity are *syntactical* in nature. They are tedious to resolve, but there are no fundamental difficulties in overcoming them. The fourth type of heterogeneity, however, is a major integration impediment. Semantic heterogeneities arise for various reasons and their homogenization needs in most cases experts from the specific domains. Consider for example the two relations shown in tables 2.1 and 2.2. The semantic heterogeneities arising when integrating these relations are:

- The attribute `capital` is used in the relation `members` to denote a person's assets and is used in the relation `cantons` for the capital city of the Canton. The attribute `name` is used for the name of a person and the name of a Canton, respectively. These heterogeneities are due to *homonyms*.

- The Canton code is `lives-in` in `members` and `code` in `cantons`, i.e., there is a *synonym* for Canton codes.

id	name	firstname	capital	lives-in
1001	Muster	Peter	$10000	ZH
1011	Sample	Sarah	$15000	GE
1023	Dupont	Pierre	$12000	GE
1030	Nowhere	John	$12000	BG

Table 2.1: Semantic heterogeneity: Relation members

code	name	capital	population	avg_income
ZH	Zurich	Zurich	1023265	Fr. 60000
GR	Grisons	Chur	70534	Fr. 55000
GE	Geneva	Geneva	1023265	Fr. 60000

Table 2.2: Semantic heterogeneity: Relation cantons

- The average income (avg_income) is encoded in a different currency from capital in members. A comparison or any other integrative task needs to transform one of the attributes.

- Relation members contains a Canton code not existent in cantons. There are three possibilities here, namely:

  1. The attribute value for this entry in members is wrong.
  2. There is an entry missing in cantons.
  3. lives-in and code do not mean exactly the same thing. lives-in might also include provinces from other countries and this one means *Bergamo, Italy* (i.e., the domain is different).

As can be seen from this constructed example, various heterogeneities are possible. In reality, however, such heterogeneities are much more concealed, often due to insufficient documentation available about the intension and extension of the component DBS and their local schemata, respectively. As will be shown later on, semantic heterogeneities are also the major impediment when integrating environmental data and therefore a major topic of this thesis. The resolution of such heterogeneities strongly depends on the expressivity of the canonical data model. It has been shown that semantically rich models such as object-oriented models provide a suitable environment both for the derivation of a federated schema and the translation and integration of local schemata (Schrefl, 1988; Härtig & Dittrich, 1992; Thieme & Siebes, 1993).

2.4 Data integration and geographical information systems

2.4.1 Overview

The use of geographical information systems as a hard- and software environment for storing, analyzing and visualizing spatial information has grown significantly in importance. This is partly due to the ability of a GIS to integrate diverse information. This ability is often cited as a *major defining attribute* of a GIS (Maguire, 1991). GIS has been called an *integrating technology* "because of the way in which it links together diverse types of information drawn from a variety of sources" (Shepherd, 1991, p. 337). The term *data integration* has been used with various meanings in the context of GIS, ranging from technical issues such as inter-conversion of raster and vector models, cartographic notions[6] to general definitions such as bringing together of spatial data from a number of sources and disciplines, from different subsystems "each employing a different technique of data input from different media" (Aybet, 1990, p. 18).

Flowerdew (1991, p. 338) sees data integration[7] as the synthesis of spatial information based on *data linkage* within a coherent data model. He distinguishes three problem domains:

1. Bringing together diverse information from a *variety* of sources, i.e., data migration.

2. Matching of supposedly similar entities in these sources.

3. Resolving *inconsistencies* across the source datasets.

That is, data integration using this definition is compatible to a large extent with the notion of the term used in the domain of classical databases discussed in section 2.3, even though the terminology is somewhat different. The following sections discuss some aspects of spatial data integration, namely integration of remote sensing data in GIS, heterogeneity issues, relevance of data quality, metadata, and system integration. The objective is to reveal the major impediments of data integration in preparation for improved concepts of an enhanced interoperability which will be discussed in chapter 3.

---

[6] A quotation often used defines data integration as "the process of making different datasets compatible with each other, so that they can reasonably be displayed on the same map and so that their relationships can sensibly be analyzed" (Rhind *et al.*, 1984).

[7] Actually, the term *information integration* is used. This is regarded as synonymous with *data integration* here.

2.4.2 GIS and remote sensing

Remote sensing is a major information source for GIS and, thus, the integration of remote sensing data within GIS is an important aspect of data integration. The data acquisition techniques used in remote sensing have improved over the past decades, providing large amounts of data of various scales. Unlike *in situ* measurements, remote sensing is able to cover large regions with spatially and temporally continuous measurements, i.e., placement of the measurement apparatus is not as critical as it is with *in situ* measurements[8]. Remote sensors are typically distinguished by the placement of their platform into *earth-borne*, *airborne* and *space-borne* sensors. They have in common that remote physical quantities can be derived, i.e., estimated, from a local measurement. Earth-borne sensors are typically mounted on the earth's surface and include:

- weather radars,
- flight control radars,
- laser interferometry systems,
- seismograph arrays and so on[9].

Space- and airborne sensors are mounted on flying platforms and called *airborne* if the platform is an airplane or balloon and *space-borne* if the platform is a satellite or space shuttle[10]. Besides their location, sensors are also differentiated into *active* and *passive* systems. Passive systems do not directly influence the objects measured and typically measure some kind of radiation emitted or back-scattered by objects, which in most cases is electro-magnetic radiation in various spectral ranges. Active systems induce energy themselves on the objects and measure effects of the interaction (e.g., radar systems).

A research initiative of the *National*[11] *Center for Geographic Information and Analysis (NCGIA)* (Star et al., 1991) focused on integration of remote sensing and GIS, covering mainly air- and space-borne earth imaging systems[12]. Within this research initiative, several impediments to integration have been identified, such as (Ehlers et al., 1991; Lunetta et al., 1991):

---

[8] However, the placements of the *sensors* is critical.

[9] Stressing the notion of remote sensing a bit further into smaller scales, a nuclear-spin-resonance tomography or X-ray apparatus could also be seen as remote sensors. Also, a telescope used for astronomy is a remote sensor, on the other far end of the scale range. Eventually, *every* measurement can be seen as a remote measurement in the sense that a "phenomenon" is never measured directly but only a kind of *interaction* with that phenomenon.

[10] Actually, a differentiation based on the distance to the earth's surface might be more sensible, e.g., up to the stratosphere *airborne* and above *space-borne*.

[11] United States.

[12] It is interesting to note that almost all literature (for a review see Hinton (1996)) on integration of remote sensing and GIS is based on aerial photographs and satellite images. Earth-borne remote sensing seems not to be present at all, even though data from earth-borne systems often pose major integration problems because of the implicit geometry in

- Different concepts of space, i.e., the much discussed duality of an *object-view* and a *field-view* (Couclelis, 1992).

- Data conversion and exchange issues (i.e., questions of data migration).

- Lack of integrated databases (i.e., GIS software with insufficient functionality for the storage and management of remote sensing data).

- Data quality management, e.g., error assessment procedures for the sampled (raw) data, errors induced by translations between data models and post-processing.

In current systems many of these impediments are overcome to some degree. Note, however, that there is no unifying concept of space yet found which meets the requirements of all domains. Without a clear understanding of what the data are that need to be represented, it is not possible to provide sensible storage and analysis techniques. Thus, the problems are clearly on a semantic level. "Converting vector data to raster data" is not – even if it is tedious sometimes – a problem on the technical level any more. It often fails, though, because there is no understanding of what, for example, a pixel in a raster model or a line in a vector model is. As will be shown in chapter 3, the provision of *sufficient semantics* is a prerequisite for successful integration.

2.4.3 Semantic heterogeneity

As with database integration, *semantic heterogeneities* are a major problem which have to be dealt with in spatial data integration. Sheth and Larson state that "semantic heterogeneity occurs when there is a disagreement about the meaning, interpretation, or intended use of the same and related data" (Sheth & Larson, 1990, p. 187). In the context of data integration within GIS, these heterogeneities are sometimes called *inconsistencies* (Shepherd, 1991, p. 342), even though this might be a bit misleading since inconsistency means not only *not in agreement* but also *contradictory*. Semantic heterogeneities *might* lead to inconsistencies but do not necessarily do so. Worboys and Deen (1991, p. 30) identify two types of semantic heterogeneity in distributed geographical databases:

Generic semantic heterogeneity This is heterogeneity based on different concepts of space or data models used, respectively, e.g., when integrating field-based databases and object-based databases.

---

such datasets. Consider for example weather radar datasets. The implicit geometry is polar (Doviak & Zrnić, 1984) in the sense that a "pixel" is given in polar coordinates by:

$$\{(r, \lambda, \phi) | r_1 \leq r \leq r_2 \wedge \lambda_1 \leq \lambda \leq \lambda_2 \wedge \phi_1 \leq \phi \leq \phi_2\} \quad (2.1)$$

That is, "pixels" are not rectangular and their volume is variable. The implicit geometry is, therefore, much more complicated than rectangular grids usually found in (processed) satellite images. However, if satellite images are being used in their raw form, the implicit geometry might be much more complicated owing to the topography of the earth's surface and propagation characteristics.

Contextual semantic heterogeneity Heterogeneity based on different semantics of the local schemata. These include heterogeneity issues from standard databases such as synonyms and homonyms.

The resolution of *generic semantic heterogeneities* needs transformation from component schemata (and datasets) into an integrated schema based on a *canonical data model*. The canonical data model can be – as seen before – one of the components' data model or a "new" data model. It has often been argued (Worboys & Deen, 1991; Couclelis, 1992; Kemp, 1993) that the object-based and field-based spatial conceptualizations expose a certain duality.   Consider for example an atomic spatial datum d consisting of a relation between a spatial reference s_i and a non-spatial entity z_i, i.e., $d = (s_i, z_i)$. The object-based approach first identifies z_i and determines s_i based on z_i's semantics. For example, z_i might be "the river Rhine" and s_i a spatial characterization which is given by the detailed semantics of z_i, i.e., scale, area-of-interest. In a field-based approach s_i would be identified first, e.g., based on some type of spatial partitioning. Then, the phenomenon z_i "at" that location s_i is identified. For example, z_i might be "1 if there is a river there and 0 otherwise", "population within s_i", "surface air temperature at s_i" and so on. Thus, the difference is basically given by the *direction* of the mapping (relation) between s_i and z_i, i.e., $s_i \mapsto z_i$ (field) or $z_i \mapsto s_i$ (object) as shown in figure 2.4.

However, for most applications there is a form which is more *convenient*. The river Rhine, for example, is much more easily expressed in an object-based approach, e.g., by using a thin and long polygon as s_i. A field-based approach would, for example, partition the area of interest into many elements s_i, e.g., square pixels corresponding to the dataset's resolution and assign each s_i to a so-called *indicator variable* stating presence or absence of the "river Rhine" feature. On the other hand, natural phenomena in general and continuous phenomena in particular are often more easily expressed as fields. Terrain height, for example, derived from some interferometry techniques, is often expressed as the *height z_i at some location s_i* and not *the location s_i of some height*. This is due to the sampling procedure which typically first determines s_i. Sometimes such phenomena are also expressed in an object-model using, for example, contour lines, which define the spatial extent s_i of z_i. Such representations, however, are mostly used for visualization purposes and not as a generic representational technique[13].

In the case of integration of data based on different data models, it has to be decided what canonical data model to use. The main objective, then, is to preserve the *expressiveness* of the component data models. This needs a

---

[13] Contour lines are frequently used, in particular for digital terrain models. This has basically two reasons. First, manual sampling procedures in photogrammetry are often based on keeping the height fixed and following the contour line, e.g., the sampling procedure first determines z_i and then s_i. Second, digital terrain models are often derived from analog maps, which contain – as a visualization – terrain models represented by contour lines with the exception of "special interest points" such as mountain peaks.

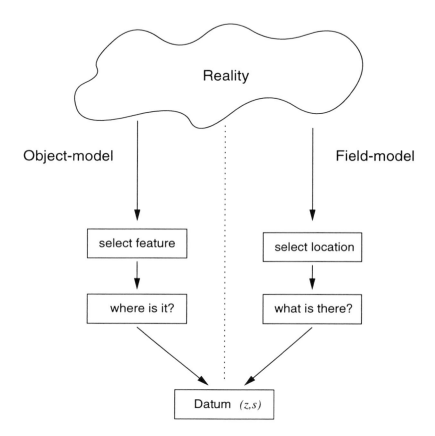

Figure 2.4: Object- and field-models

canonical model which is able to represent both spatial objects and fields in a coherent way[14]. The main difficulty lies in the existence of *implicit semantics* within the various data models. For example, topological relationships are often managed on the level of the data model. In a field-based model the topological relationships are restricted (in most cases) to classical neighborhood relations within the spatial partition, i.e., relationships between non-neighboring spatial references s_i are either not present or implied, whereas such relationships can be explicitly modeled in an object-based approach by appropriate geometry elements (e.g., line strings and their end-points).

Contextual semantic heterogeneities correspond to the heterogeneities in standard databases shown before, i.e., issues of synonyms, homonyms, mismatching definition domains and general data quality issues. Some of the conflicts particular to spatial databases are discussed in Parent *et al.* (1996) and Laurini (1996). Unlike standard databases, however, it does not usually cause major problems to identify the *common elements* on an *intensional* level: in most cases the linkage happens through the spatial attributes[15]. This is mainly due to nature's locality. Most physical relationships expose a certain *local* character. For example, their mathematical models are based in most cases on partial *differential* equations, i.e., a quantity depends on *local* spatial and temporal changes, or, in other words, on local interaction as opposed to remote interaction. This has been called the "first law of geography" by Tobler (1979).

It is natural, therefore, to base a linkage between two component DBS or datasets[16] on their spatial attributes which are easily identified. Consider two spatial data $a = (z_a, s_a)$ and $b = (z_b, s_b)$ where the non-spatial attribute z_a is related to a spatial reference s_a and likewise z_b relates to s_b. We will first assume that the relation is an injection, i.e., it is invertible. This means, that we can conclude from $z_a \mapsto s_a$ that $s_a \mapsto z_a$. The linkage happens as mentioned before on the spatial attributes, that is:

$$\begin{array}{l} z_a \mapsto s_a \\ s_a = s_b \quad \longrightarrow \quad z_a \mapsto z_b \\ s_b \mapsto z_b \end{array} \qquad (2.2)$$

This procedure is generally called *spatial join* or *polygon overlay* if s_a and s_b are polygons.

In reality, however, the situation is much more complicated than shown in equation (2.2). First, there are rarely identical spatial references $s_a = s_b$. In the case of polygons, for example, the two polygons s_a and s_b which are not

---

[14]Such a model is discussed in chapter 3, section 3.4.
[15]Geographic databases are not different from standard databases if the linkage is based on *non-spatial* attributes.
[16]We will make no distinction between "dataset" and "database", here. This is introduced in chapter 3.

disjoint are first intersected. This yields a partitioning of both polygons with $s_a = \bigcup_{i=1}^{n} s_a^{(i)}$ and $s_b = \bigcup_{i=1}^{m} s_b^{(i)}$. Within this partitioning there are pairs of numbers (i,j) identifying equal references $s_a^{(i)} = s_b^{(j)}$ which will actually allow a z_a to be related to a z_b as given in (2.2). Semantical heterogeneities are similar to those in classical databases:

- The relations $z_a \mapsto s_a$ and $z_b \mapsto s_b$ might not be *injective*, i.e, situations of homonyms and synonyms. Spatially speaking this means that a spatial reference s maps to several entities z_i or that an *entity* is referenced by several spatial references s_i.

- The relations $z_a \mapsto s_a$ and $z_b \mapsto s_b$ might be not *surjective*, i.e., there might be spatial references s that are not related to any entity z.

In most cases, however, the major incompatibilities or heterogeneities stem from different spatial domains, i.e., different sets of spatial references. There are techniques to *derive* a new domain which is compatible to both datasets by means of, for instance, polygon overlay, for cases where the spatial references are not disjoint. Often the spatial references are *points* in space and time, i.e., there is certainly no overlap unless two points are exactly equal. In a classical environment this would disable any integration attempt because there is no *extensional* overlap, i.e., there are no elements that can be identified as being related to each other. But there is a fundamental difference in geographical databases which is based on the locality mentioned above:

- Given a datum (z_a, s_a) it is, for example, sometimes possible to sensibly *predict* or *estimate* similar entities at nearby locations. Especially when z_a is describing a physical quantity, it is possible to estimate *values at unsampled locations* by means of interpolation or extrapolation .

- The matching of spatial references does not necessarily need to be based on exact *equality* to be meaningful. In many cases an approximate match is useful as well, i.e., allowing a certain tolerance in the match. This is possible, of course, because space and time can be modeled as a *metric space* which allows concepts such as *distance* to be used.

The next section will discuss some additional heterogeneity aspects which are related to the notion of *data quality* and which have an important impact on integration success and reliability, respectively.

2.4.4 Data quality

Data quality issues are an important type of semantical heterogeneity and are important both for the integration process and for the quality of the integrated databases. For integration, Shepherd (1991, p. 342) identifies the following data quality issues[17]:

---
[17] These issues are actually subsumed as *data inconsistencies*.

- errors in measurement of survey methods,
- vagueness or imprecision in definitions (spatial or thematic),
- fuzziness in spatial objects,
- variations in the use of terminology and nomenclature.

Data quality relates to data integration in various areas. First, there are many heterogeneities which are *due* to quality issues and which have to be resolved in integration. Second, integration often leads to some translations of the source dataset or component DBS, respectively. It is necessary that uncertainties in source data are properly propagated to integrated datasets. Third, data integration brings data together from *various* sources, e.g., collected by various organizations. Data quality has to be documented both in the source data as well as in the integrated data.

There has been extensive research on the effects of uncertainties in component databases and their propagation into an integrated database (Veregin, 1991; Chrisman, 1991b; Maffini *et al.*, 1991; Lodwick, 1991; Lanter, 1990; Hunter, 1992). Much of this research is related to general quality management of spatial data as reviewed in (Chrisman, 1991a) and (Goodchild, 1991). Assessment, analysis and documentation are often based on a typology introduced by the *Spatial Data Transfer Standard (SDTS)* (Morrison, 1988) ¡indexSDTS which consists of the following data quality elements:

- *Lineage* (Clarke & Clark, 1995) refers to the history of a dataset and describes the source material from which the data were derived and the methods of derivation. In the context of integration this refers to the documentation of the component DBS and the integration procedures (e.g., translations).

- *Positional accuracy* (Drummond, 1995) refers to the compliance to a spatial registration standard, i.e., accuracy of the spatial attributes s_i of a dataset.

- *Attribute accuracy* (Goodchild, 1995) is quality information on the *thematic* component, i.e., non-spatial attributes z_i of a dataset.

- *Completeness* (Brassel *et al.*, 1995) identifies the exhaustiveness of a dataset in comparison with its abstraction of reality, i.e., if all real-world entities that have a conceptualization are represented. This has a great importance for data integration since integrated datasets are *at most* as complete as the source datasets, i.e., their completeness is typically lower.

- *Logical consistency* (Kainz, 1995) describes the compatibility of a datum with other data in the dataset. This affects both spatial and non-spatial attributes. Inconsistencies within datasets and between datasets are clearly an issue for data integration.

Besides these five elements, two others were considered in the *Commission on Spatial Data Quality* of the *International Cartographic Association (ICA)*:

- *Semantic accuracy* (Salgé, 1995) somehow complements the completeness element in that it identifies the accuracy of the semantic content of a dataset which in turn defines the abstract universe against which completeness is evaluated.

- *Temporal information* (Guptill, 1995) was included in the five quality elements from SDTS as *currency*, i.e., giving information about how current a datum in a dataset is. This has been expanded to include general temporal *quality* information, with respect to *observation time* and *database time*[18]. The *event time*, i.e., a datum's temporal coordinate, should be dealt with as *positional accuracy*.

These data quality issues are[19] an integral part of a dataset's documentation as discussed in the next section. Because analyses based on spatial data are communicated in most cases through *maps* and other graphical representations, data quality must also be included in such visualizations. There has been some research in that area, e.g., (Beard *et al.*, 1991; MacEachren, 1992; Beard & Mackaness, 1993; Mackaness & Beard, 1993), but its application is yet far from being operational. Also, the inclusion of quality information for decision making – which is the main rationale for making analyses anyhow – is not at all routine as shown in (Stanek & Frank, 1993). This is somewhat astonishing, since the need for the inclusion of data quality information on a more detailed level was already proposed in (Chrisman, 1983).

2.4.5 Metadata

Metadata management has become a major concern in parallel with the increased use of spatial databases. There have been many attempts to give a concise definition of *metadata*. Lenz (1994, p. 160) defines metadata as "data about data which describe the (data) [...] on the semantic, structural, statistical and physical level." Other authors generally define metadata as "information which makes data useful" (Bretherton & Singley, 1994, p. 166) or "information someone forgot to include in a dataset" (Dutton, 1996). In the context of statistical databases the concept of metadata has been expanded to include not only a description about the meaning of the numbers in a dataset, but also a description of the conceptualization of reality (van den Berg & de Feber, 1992, p. 296). As can be seen from these definitions, *metadata* is an overloaded term and means different things in different contexts

---

[18] Sometimes also called *transaction time* and referring to the time at which the datum was inserted, updated or deleted in a database.

[19] One ought to write "should be" here since there are still many spatial databases around *with no quality information at all*. No quality information is – in a similar manner to Paul Valéry's statement "Le mélange du vrai et du faux est plus faux que le faux" – often worse than a dataset of bad quality.

(Bretherton & Singley, 1994). In database theory, metadata is a technical term that means *information about a database schema* or *data dictionary*, i.e., description of all classes and their attributes. In the context of spatial databases, this has been somewhat extended to include any "additional" information needed to make a dataset useful. The *Metadata Content Standard* of the US Federal Geographic Data Committee (FGDC) (FGDC, 1994) provides a *structurization* of additional information which might make a dataset useful. The standard defines many attributes and their value domains which allow structured and comparable documentation of spatial data sets. The attributes are part of a hierarchy covering various aspects such as *data quality*, *data sources* and *references*.

It has been often argued where the "data" end and "metadata" start. For example, Chrisman (1994) notes that metadata about spatial data are spatial data themselves. This is clearly true since a description of a dataset will typically contain some information about the *extent* of the data contained therein. The confusion also arises from various uses of metadata. There are three major categories of metadata use identified in (Blott & Včkovski, 1995) with large differences in the level of detail:

Production and quality control The production of large spatial datasets sometimes involves many persons and requires many years for their completion. As in every engineering task, detailed metadata are essential for production. In such a project, the metadata might be huge and would contain, for example, work lists of all persons that created or modified parts of the dataset, agreements with sub-contractors, detailed specification of the sampling procedures and everything else needed as project documentation. In some cases the metadata might be even larger in volume than the "actual" data. Such metadata are especially important for quality assurance and control, e.g., (Gilgen & Steiger, 1992).

Data exchange Metadata play a very important role in data exchange. The metadata accompanying a dataset are a key to successful exploitation of that data. The necessity to provide *contracts* as discussed in section 2.1 for spatial data exchange also applies to metadata. The FGDC Metadata Content Standard is an example of such a standardization[20]. In most cases, metadata are encoded as a free-form text, possibly adhering in structure to a content standard. Availability of sufficient metadata is also a key factor for a successful integration, especially if *semantical* heterogeneities are to be resolved. The level of detail of such metadata is typically less than that of metadata used for production purposes and can vary from a few pages of text for small datasets to a few hundreds of pages.

Data directories In a larger context, metadata are used not only as a data *dictionary* (i.e., defining the items *within* a dataset) but also as data

---

[20]This standard standardizes the *content* of a metadata description and not its encoding.

directory, providing information about available datasets. A metadatabase is a collection of dataset descriptions which is used to locate datasets within organizations or data libraries. Such metadata are usually less detailed than the previous two examples and contain information on the level of detail found in library catalogs. Another difference to the uses described above is that such dataset descriptions are often complete in the sense that they can exist without the dataset they are referring to and they can be independently exchanged, e.g., between libraries.

These three usages of metadata seem to be very different at first glance. However, the main difference lies in the level of *abstraction* or *generalization* applied to the dataset. From that point of view the border between *data* and *metadata* begins to disappear: metadata can be seen as a generalization of a dataset to the desired level of detail. Consider for example a book containing several chapters, a table of contents, an abstract and a title. The chapter titles are generalizations of the chapter contents. The table of contents and the abstract are two *different*[21] generalizations of the book's content on a similar level of detail and the title is a generalization on the lowest level of detail. A library catalog would typically contain only some of the generalizations on a low level of detail, i.e., author name, title, publisher and so on. A recommendation of a book sent to a friend may include a copy of the abstract. The author writing the book would collect substantially more information than is necessary for the production of the book.

The need to "add metadata" to a dataset basically means that it is necessary to provide data at *various generalization levels*[22]. As will be also shown in a case study on the language POSTSCRIPT (section 6.2), this is a prerequisite for successful data exchange and integration.

2.4.6 System integration

System integration is an interoperability issue within geographical systems as well. This refers to *bringing together* previously separated systems. In most cases this means interoperability between separate applications and can be reduced to a data *migration* or data *integration* problem. In environmental applications, especially, it is often necessary to provide an integration framework for highly specialized applications. Such a framework "provides scientific investigators with a unified computational environment and easy access to a broad range of modeling tools" (Smith *et al.*, 1995, p. 127). Such an integration of applications is often referred to as *cooperative systems* or – in a more commercial context – *workflow management*. The cooperation is between multiple components or agents which constitute a complex computational environment (Alonso & Abbadi, 1994, p. 84). Such agents might

---

[21] It is important to note that there are always various valid generalizations with the same level of detail possible.
[22] Generalization is not meant in a cartographic sense here, but it does include cartographic aspects.

represent data sources (e.g., remote sensors, ground observations), repositories (e.g., database management systems), researchers (e.g., hydrologists, geologists, geographers, computer scientists) and tools (e.g., image processing, statistical analysis, numerical simulation).

The basis of such a cooperative system is – again – a coordination layer or *contract* between the components or agents of the system. As will be seen in the next chapter, a contract which is expressive enough can both serve for data integration and system integration. In most applications, the cooperation is based on the exchange of data only and, therefore, a contract is based on data formats and the like, e.g., (Djokic, 1996; Maxwell & Costanza, 1996).

System integration on the lowest level simply refers to the ability to share hardware resources, e.g., run on the same hardware platform, use the same output devices. It has been a major concern in many organizations having large legacy applications to achieve interoperability, *at least* on the user interface level (Gust, 1990). This means, for example, that it should be possible to use different applications from the same computer monitor and keyboard. In asset management companies, brokers often still have multiple monitors in front of them. The different monitors are not there to provide more simultaneous information to the broker, but are there because the various systems are not interoperable on the level of the human/computer interaction.

Albrecht (1995) discusses a system integration approach called *virtual GIS*. The main idea thereby is to shield the user from low-level details and commands of a GIS and provide the ability to concentrate on specific *tasks* rather than data management and GIS operating. The strategy followed is based on an identification of *elementary* GIS functions and the relations between these operations. A task is built up from a set of such generic functions with a specific goal in mind.

2.4.7 Environmental data management

The management of environmental data is a classical application area of GIS. However, in many cases the GIS is merely used as digital mapping system (Wheeler, 1993, p. 1498), supported by modeling applications external to the GIS. This is partly due to some special characteristics of environmental data leading to semantic heterogeneities which cannot be easily overcome. The variety of requirements for the management of environmental data and, therefore, their heterogeneity has been reported many times and is reviewed, e.g., in (Kapetanios, 1994). Here, a short summary of the major particularities of environmental data is given:

Large volumes Data can be large in size. Data from automatic measurement equipment such as remote sensors continuously produce huge data volumes. This is challenging not only for integration and interoperability approaches, but also for mere mass-storage management.

Multi-disciplinarity Environmental data are often used and produced in various scientific domains. The investigation of environmental phenomena is generally a highly interdisciplinary task. In such settings, semantic heterogeneities are very likely to occur, out of different "common sense" within disciplines[23]. In addition to these semantic issues, there are also syntactic issues which arise due to conflicts in established data formats and software collections within the individual scientific domains.

Multi-dimensionality Environmental applications often go beyond a flat and static world as is used as a basic earth model in many GIS. This means that many datasets are referenced in a four-dimensional coordinate system (Mason *et al.*, 1994). In particular, the management of a *continuous* and *dense* temporal dimension with a geometry comparable to "classical" spatial dimensions is not yet possible in many GIS[24].

Field models Environmental data are almost always present as *fields*, i.e., the spatio-temporal location s refers to a thematic value z as was shown in figure 2.4 on page 24. The set of spatio-temporal references is most often induced by external factors such as sensor characteristics, sampling strategies, cost of apparatus and not by the phenomenon investigated[25].

Locality The locality or spatio-temporal autocorrelation of environmental data as described on page 26 often allows new values to be estimated from available values. This is very often needed in environmental applications (Hinterberger *et al.*, 1994) in order to resolve heterogeneities, like mismatching spatial references in a spatial join.

Data quality Data quality issues are also of high importance in environmental data management, particularly when it is used to assess a particular dataset's fitness for use in a specific application. In an interdisciplinary setup, especially, it is important to communicate the data quality across discipline boundaries, i.e., *interdisciplinary communication* as it is put forward by Foresman and McGwire (1995, p. 16).

Metadata The importance of metadata is given both in a multi-disciplinary setup and within single disciplines. It is necessary to document and

---

[23]The OGIS approach presented in section 3.4 on page 46 uses the term *information communities* to describe such diverse domains.
[24]Time and GIS has been an often discussed topic and various "types" of time have been identified, e.g., (Al-Taha & Barrera, 1990; Worboys, 1992; Qiu *et al.*, 1992; Özsoyoğlu & Snodgrass, 1995). In the context of environmental data management, so-called *event time* is of interest. Event time is continuous and represents the time of the observations. The type of geometry on time has often been a topic itself. It has been argued, for example, that it is directed and needs other geometry concepts. However, there is – except for the second law of thermodynamics – not much physical evidence of a *direction* of time (Ridley, 1995, p. 67).
[25]Although, a good *sampling strategy* does of course consider the known characteristics of the phenomenon.

communicate as much information as possible in order to allow an appropriate use of datasets. Moreover, metadata are needed – as discussed above – to locate datasets in data directories and data library catalogs.

Numbers As opposed to, for example, commercial or administrative information, scientific data in general and environmental data in particular consist in majority of *numbers* and mostly *rough numbers*. In most cases, these numbers are also meant to carry a *unit*, i.e., they are *not dimensionless* numbers. When integrating such data – especially when the data are crossing discipline boundaries – it is very important to include information about roughness and units of number values.

2.5 Review

This section has presented in rough outline issues of interoperability and data integration in the context of GIS. The inventory has shown a variety of requirements and possible problems, mainly being variants of syntactic and semantic heterogeneities. It is the topic of the next chapter to propose a strategy for (data) interoperability based on an analysis of the major impediments.

The range of applications of GIS is much larger than environmental applications discussed in the last part of this chapter. Data integration and interoperability problems in other application areas, such as for example management of communication networks, are considerably different at first glance, also because an object-view of the earth is adopted, as opposed to field-views as used in environmental applications. However, it is the objective of the remainder of this thesis to address the integration problem on a more general level. This is attempted by using field models *as an example*. It might sound contradictory to say that the problem is addressed on a general level on the one hand and then deal with it only within one branch on the other. However, the important point is that the treatment is *an example* and the analogy to other branches should be given implicitly in most cases. One reason why field models where chosen, as was said before, related to the overall background of this work. In addition, most other studies on interoperabiliy in GIS are concerned with other application domains such as facilities management. That is, after all, where the interesting markets for software vendors are. Therefore, the scope of this thesis is more on environmental applications.

CHAPTER THREE

Virtual Data Set

3.1 Overview

This chapter collects the interoperability issues discussed in the previous chapter and identifies some fundamental interoperability impediments of spatial data. Spatial data interoperability or *geodata interoperability* is referred to as "the ability to freely exchange all kinds of spatial information about the Earth and about objects and phenomena on, above, and below the Earth's surface; and to cooperatively, over networks, run software capable of manipulating such information" (Buehler & McKee, 1996, p. x). This means that the two key elements are:

- information exchange

- cooperative and distributed data management

The latter objective is actually a special form of information exchange. Cooperative and distributed data management means that there is *bi-* or even *multi-directional* information exchange in the sense that, for example, a data user might also *update* a remote data source in addition to simple querying.

Based on the identification of the key geodata interoperability difficulties, some approaches for an improvement are analyzed. These improvements are addressed, for example, by the *Open Geodata Interoperability Specification (OGIS)* discussed in section 3.4. The concept proposed in this thesis is called the *Virtual Data Set (VDS)* and can be seen as a subset of OGIS. The VDS is presented in section 3.5 and is related to general trends in current software design.

3.2 Interoperability impediments

3.2.1 Syntactical diversity

Simple encoding differences are still a major problem area within the exchange of geodata. These issues, however, are addressed by many *format* standardization efforts as discussed in section 2.2 of chapter 2. Syntactical issues are especially important in the context of quality management. The resolution of syntactical diversities currently often results in tedious automatic and manual format conversions, involving many steps and software tools. Such procedures are likely to affect the data quality for various reasons, such as:

- The target format may have a lower expressivity than the source format.

- Some *intermediary formats* may have a lower expressivity than the source format. This case is dangerous because the usage of an intermediary could be hidden in some software tool.

- It is likely that such *tedious* conversions are not assigned to well-trained personnel.

- Issues which are actually due to semantical heterogeneities might be interpreted as syntactical problems. A typical case are *homonyms*.

- Syntactical heterogeneities might not be detected. For example, different text encoding or byte ordering in formats which do not define these issues[1].

In many cases, many of the syntactical incompatibilities are also found at lower levels such as media compatibility issues, e.g., "how can I read tape X on my computer Y?". Fortunately, most such problems are nowadays addressed by the operating system.

A source of these problems is often the lack of explicit specifications. Many format specifications are based – sometimes unconsciously – on implicit assumptions such as measurement units used and low-level encoding schemata.

3.2.2 Semantic diversity of geographic information

Geographic information is inherently complex. As discussed in section 2.4.3 of chapter 2, the canonical data model needed for interoperability needs to

---

[1] Text encoding issues are frequent sources of problems. Consider, for example, a data format allowing some multi-line text blocks to be included. Often, it is only defined which *character encoding* to use for the text (e.g., ASCII). The line encoding mechanism is implied in most cases, assuming some special end-of-line characters to mark the end of a text line. It is likely that implementations of such a format definition will use the *platform specific* end-of-line characters. For example, Macintosh systems use ASCII character CR, Unix systems character LF and MS-DOS/Windows systems the sequence CR-LF. Text blocks transferred within such a format between, e.g., a Macintosh system and a Unix system, are likely to be interpreted differently on the target system.

express the complexity. The complexity is multiplied by the number of cooperating information communities because of semantic diversities. Many integration approaches fail because of these heterogeneities. Therefore, it is of paramount importance to address semantic heterogeneities in any interoperability approach.

As with the syntactical issues discussed above, semantic heterogeneities are often *not detected* because of the lack of sufficient specifications, i.e., definition of the semantics. Consider, for example, a spatial data set describing the river *Rhine* by a sequence of straight line segments. The *resolution* of this dataset is important for the evaluation of the fitness-of-use of the dataset for a specific application. With such a line segment, or actually, a set of connected points, it is not trivial to determine the dataset's resolution. Are the points the only measurements known and the line between them a bold estimation? Or, were the points selected in a manner which guarantees that a straight line drawn between the points generates at most a specified distance error to the "true" location of the Rhine? Or, is it known that the points on the line are as trustworthy as the end points of the line[2]? The semantics of what the points and the line between them mean needs to be very clear to anyone using such a dataset. This is no problem as long as data exchange happens within an information community with a well defined knowledge or if data can be understood by common sense and implicit assumptions. However, such implicit assumptions can be *very* dangerous.

In many cases, there is no other way to find and resolve such semantic problems other than through human reasoning. There is a trade-off here between providing ease of use and human involvement. It suggests the use of concepts from decision-support systems to aid the resolution of semantic heterogeneities.

Another problem associated with semantic heterogeneities and the complexity of geographic information is the fact that the future use of any dataset usually cannot be foreseen. This is especially true in scientific environments where insight is often gained by explicitly using methods and usages which were not initially foreseen. This means in turn that it is not possible to solve semantic heterogeneities by defining a one-and-only data model, implementing a respective schema, and translating all data into that schema as it is done, for example, in the Sequoia 2000 project (Stonebraker, 1994). There will be a future use with requirements that do *not* fit that model and schema and will expose semantic heterogeneities. An example are data values which can be derived in a simple manner from other data values, such as spatially interpolated values, statistical aggregations, or unit conversions. The variety here is too large to be foreseen and statically represented (e.g., by pre-calculating all possible aggregations). An application using such derived values would typically identify the context which defines the way derived values are needed

---

[2]This is often the case with human-made structures, for example, when exchanging data in *Computer Aided Design*. There, the "truth" of any of the points on the line is typically not less than that of the end points.

such as type of aggregation, location at which interpolation is needed and so on. This does *not* mean, however, that the derivation belongs exclusively to the problem domain of the application! Rather, it is a *property* of the dataset used, only that a derivation *can* be performed when the application context is known. In other words, this means that in order to resolve these issues, *rules* for the derivation need to be available.

3.2.3 Diverse information communities

Interoperability is not only an issue between systems but also between people. Many of the semantical heterogeneities are due to different essential models in various application domains, that is, different concepts of reality and different languages used to describe these concepts. Information communities are defined in OGIS (Buehler & McKee, 1996, p. 54) as

> "a collection of people (a government agency or group of agencies, a profession, a group of researchers in the same discipline, corporate partners cooperating on a project, etc.) who, at least part of the time, share a common digital geographic information language and share common spatial feature definitions. This implies a common world view as well as common abstractions, feature representations, and metadata. The feature collections that conform to the information community's standard language, definitions, and representations belong to that information community."

The existence of diverse information communities is without a doubt one of the major reasons for creating semantic heterogeneities of all kinds. In many cases, it is and will be necessary to bridge the differences between such information communities, i.e., data migration and integration does not happen entirely *within* an information community. Geodata interoperability needs to address the resolution of heterogeneities between such information communities and provide mechanisms that can be used to "map" the semantics between different information communities. However, this is a very difficult topic and it seems not to be very clear what such a mapping should look like. The OGIS Information Communities Model, for example, relies "on the use of special registries that contain manually derived semantic models which enable a mapping of terms and/or definitions from one information community to another" (Buehler & McKee, 1996, p. 56).

Traditionally, the heterogeneities between information communities are addressed by a suitable set of metadata which makes explicit as much information as possible. Heterogeneities can be detected and resolved because all terms and definitions are explicitly listed and can be compared with those that are in use in the "target" information community. Such a procedure, however, is error-prone since it demands both an understanding of the definitions and terms given in the metadata of a dataset *and* a proper understanding of the corresponding definitions in the "target" information community. It is not even uncommon that such definitions are not exactly known *within* an information community. Assumptions and (essential) models may be used without being aware of them. In such cases, heterogeneities cannot be detected, of course.

3.2.4 Market forces

Most spatial data handling happens in commercial areas such as facility management. An interoperability approach can, therefore, be successful only if it is also commercially successful, that is, if the technology is accepted by vendors and customers. However, market forces can sometimes be inhibitors in standardization processes:

- Standardization also often means *openness* and makes it difficult to hide proprietary technology.

- Standardization produces more competition because different vendors can be considered for extensions to existing interoperable systems.

- Standardization can disable *innovation* because it does not allow new technologies to be adopted.

- Standardization processes are sometimes slow and do not meet the *time-to-market* needs of commercial developments.

- A useful standard needs a certain degree of *persistence* to be useful. This is somewhat contrary to the high turnover rates wanted by vendors.

There are, however, also market forces *supporting* standardization. These are, on the one hand, the customers. Interoperable systems meet the customers' needs in that systems or components from various vendors can be chosen. This was one of the important factors for success of MS-DOS-based personal computers in the eighties: the hardware specification of the "standard IBM PC" was open and there was an industry emerging which was producing compatible peripheral devices and also compatible personal computers (so-called "IBM compatibles"). A customer could buy a computer from one vendor and a memory extension from a different vendor and the memory components did fit into the computer. This produced both a large palette of products and lowered the prices because of greater competition. Also, the *quality* of the products might be better under such circumstances (even though this was not always the case). There were third-party products also available for other computing platforms before, but these were mostly peripheral devices. The important change was that the *central processing unit* was "replaceable" by products from other vendors[3].

---

[3]However, the *operating system* and the *microprocessor* were, for a long time, available only from a single vendor, namely Microsoft and Intel, respectively. This gave these two enterprises a dominant position on the market.

On the other hand, standardization and interoperability also provide a business opportunity for small companies and even individuals. It is not necessary any more to build comprehensive systems in order to be able to sell them. Small components which do one task and do that very well can also be sold because they can be integrated into a larger system. Also, customers can trust a *small* company for *small* products. The central requirement here is the ability to provide support and maintenance for the product and this can be guaranteed for a small product (which is also replaceable in the worst case); whereas for a large and expensive system, a customer usually needs the security given by a large vendor. The commercial rise of the Internet, for example, provided business opportunities for many small companies selling specialized software products which work on top of Internet protocols, e.g., by writing a *client* component of a client/server system without having to provide a corresponding *server* component. Standardization processes are, however, also sometimes problematic for small and medium enterprises (SMEs). Large vendors are able to actively participate in and influence such a process. Several individuals can be exclusively assigned to work on and with the standardization. While SMEs are usually welcome in such processes by, for example, low membership fees in the corresponding standardization organization, it is typically not possible for a SME to follow the process with the same intensity. Therefore, SMEs play, in most cases, a *reactive* role in that products compatible with the standard are introduced at a later point in time than products of large vendors.

Finally, it has to be noted that the *complexity* of current computing systems *requires* standardization. A standard provides – among other things – a break-down of the complexity, and more importantly, an *agreed* break-down. Even if there is no externally imposed standard, it is necessary to provide "internal" standards when designing complex systems, creating syntactically and semantically well-defined interfaces between systems, problem domains, development groups and enterprises. This also allows vendors to enter into strategic alliances with other vendors which sell complementary products. Such coalitions can provide the necessary market advantage when proprietary solutions have become obsolete.

3.3 Approaches for improvements

3.3.1 Data definition languages

Many strategies for migration and integration lack the flexibility to adapt to usages, formats and systems that have not been foreseen at the design stage. In the context of data migration, there are approaches such as INTERLIS (INTERLIS, 1990) and EXPRESS (Bartelme, 1995, p. 295) which address the problem of static and inflexible data formats by introducing a data definition language. INTERLIS is a data exchange mechanism used in the Swiss

Official Survey. It is based on a *description* of the transferable data model[4] which serves as a contract between the data producer and data consumer. Furthermore, it allows the specification of a data *format* which means in the INTERLIS context an encoding of the data model, i.e., a data structure. IN-TERLIS is based on a relational data model with some extensions for spatial data. Listing 3.1 shows an example of an INTERLIS data definition. Using such a data definition, an import and export translator of a GIS can automatically derive the exchange format by automatically interpreting the data description. The data description is part of the data exchanged and can be seen as a structured part of the metadata.

A similar example is EXPRESS, which is the data definition language used in the international standardization attempt by the International Standardization Organization (ISO) and Comité Européen der Normalisation (CEN) in their technical committee 211/287. EXPRESS is a generic data definition language and goes beyond the possibilities of INTERLIS but lacks the specific spatial constructs that are available in INTERLIS. EXPRESS enhances the possibilities of INTERLIS by providing support for functional elements. It is, for example, possible to declare attributes as *derivable* and define the rules to derive the attribute's value, using control flow statements available in EXPRESS. These can also be used to define more complicated relationships between entities, consistency constraints, and so on. Listing 3.2 shows a simple example of an EXPRESS data definition.

The approaches using a data definition language are very flexible and impose almost no limitations in terms of the expressivity of the data exchange formats. However, a great deal of complexity is introduced both for the data consumer and data producer if that flexibility is required. If there is no fixed data definition, that is, if the data definition may change for every data transfer, then a data consumer needs to *understand* the data definition. It is not possible to have a "hard coded" version, but the data definition "program" needs to be interpreted. As will be seen in the POSTSCRIPT case study in chapter 6 this is a severe interoperability impediment which is based on the non-trivial task of writing an INTERLIS or EXPRESS interpreter, respectively, and making use of such a compilation.

3.3.2 Object-oriented models

Object-oriented[5] concepts and their terminology are not always very well defined as Worboys (1994) notes. Here, a definition given in (Firesmith & Eykholt, 1995, p. 294) will be used:

---

[4]INTERLIS uses the term "data model" which might be somewhat confusing since it relates to what is called "schema" elsewhere.

[5]Object does *not* refer to objects in the object/ field duality of spatial models. Of course, entities identified in space can be modeled by object-oriented principles and the object identification is often straightforward. However, this is also true for fields, only that the objects might be a bit more abstract (Worboys, 1994, p. 387).

Listing 3.1 Example of INTERLIS data definition, taken from (INTERLIS, 1990)

```
TRANSFER example;

DOMAIN
  Lcoord = COORD2   480000.00    60000.00
                    850000,00   320000.00;

MODEL example
  TOPIC ground cover =

    TABLE grsurfaces =
      type: (buildings, fixed, humusrich, water, stocked,
      vegetationless);
      form: AREA WITH (STRAIGHTS, ARCS, CIRCLES) VERTEX Lcoord;
    NO IDENT
      !! search via geometry or buildings
    END grsurfaces;

    TABLE buildings =
      assno: TEXT*6;
      surface: -> grsurfaces // type = buildings // ;
    IDENT
      - assno    !! assno adoption is unambiguous.
      - surface !! buildings are allocated an exact surface
    END buildings;
  END ground cover.
END

FORMAT FREE;

CODE
  FONT = IS08;
  BLANK = DEFAULT, UNDEFINED = DEFAULT, CONTINUE = DEFAULT;
  TID = ANY;
END.
```

Listing 3.2 Example of EXPRESS data definition

```
SCHEMA example;

ENTITY person
  SUPERTYPE OF (ONEOF (man, woman));
  firstname       : STRING;
  lastname        : STRING;
  middle_initial  : OPTIONAL STRING;
  birthdate       : DATE;
  children        : SET [0:?] OF PERSON;
  license_no      : OPTIONAL STRING
  owns            : SET [0:?] house;
DERIVE
  age             : INTEGER := years(birthdate);
INVERSE
  parents         : SET [0:2] OF person FOR children;
WHERE
  valid_license   : (EXISTS license_no AND age > 18) XOR
                    NOT EXIST license_no;
END_ENTITY;

ENTITY house
  on_parcel       : parcelnumber;
  address         : STRING;
  value           : NUMBER;
  owner           : OPTIONAL person;
  public_office   : OPIIONAL STRING;
WHERE
  usage           : EXIST (owner XOR EXIST public_office)
END_ENTITY

END_SCHEMA;
```

"[Object-orientation is] the paradigm that uses objects with identity that encapsulate properties and operations, message passing, classes, inheritance, polymorphism and dynamic binding to develop solutions that model problem domains."

We will not go into details of object-orientation here but refer to standard literature about object-oriented techniques, e.g., (Booch, 1991; Cook & Daniels, 1994). The notion of *encapsulation of properties and operations* is important here. Or, as Worboys (Worboys, 1994) puts it:

object = state + functionality.

In the context of interoperability, this means that the complexity can be reduced while keeping high flexibility and expressivity. The complexity reduction is given by encapsulation. Much of an object's complexity is *hidden* for the object user, and the object is typically only accessed by sending and receiving a set of messages to and from the object (which is synonymous with "invoke operation" or "call methods"). Flexibility and expressivity are given by the functionality embedded in an object. The object's behavior is only limited by how the functionality can be expressed. This is especially important for *derived* values. Object-oriented models allow such values to be defined by methods which do not directly return the object's state but return instead some derived values. Derived values have been used in scientific databases for various purposes, such as conversion of measurement units, aggregations and so on (Smith & Krishnamurthy, 1992). The derivations are often given as *data-driven rules* in the database system. Data-driven rules are primitives in a database system that can be inserted, modified, or deleted and which are activated when a database's data items are modified (Gal & Etzion, 1995). Such data-driven rules are often used to maintain derived values in addition to other activities, such as enforcing integrity constraints and triggering external operations. Database systems having extensive support for such *event–condition–action* rules are usually called *active* databases (Chakravarthy, 1995; Hanson, 1996). Using data-driven rules, derived values can always be kept up to date. However, the derivations in such systems happen when the *source data* change and not when the derived values are requested. This works as long as the derived value does not depend on the context, i.e., there are no user-definable parameters in a query for derived values. In the latter case, there is no way around deriving values on request, unless the range of parameters is small and it is feasible to have "subscripted" derived values as in (Pfaltz & French, 1990). In an object-oriented model, it is left open to the object implementation (and hidden from the object user) to decide whether derived values are computed in advance (e.g., triggered by a state change) or on request.

Object-oriented techniques are therefore very useful to achieve interoperability. It is, however, not always simple to employ object-oriented techniques when *migrating* data. While the transfer of *state* is exactly what is done, usually, in data migration, the transfer of *functionality* is not straight forward.

Systems such as Java, discussed in chapter 7, provide a basis for full object exchange, i.e., state *and* functionality.

3.3.3 Semantic accuracy

Many of the problems discussed before relate to *semantic accuracy*. Salgé (1995, p. 145) defines this as "[a measure for] the semantic distance between (...) objects and the perceived reality". That is, it somehow measures the goodness of the *specification* of a dataset. The lower the semantic accuracy is, the more semantic heterogeneities have to be expected[6]. How can semantic accuracy be addressed? Semantic accuracy is a very broad topic and covers almost all aspects of a dataset. Consequently, every action and procedure within a dataset might have an influence on its semantic accuracy. There are, however, some points which might help to give an overall increase of semantic accuracy, such as:

- Make all information explicit, i.e., do not rely on implicit information. This means a sort of *inner completeness*.

- Provide accurate *essential models* as a basis of a dataset's specification.

- Provide as much information about the data as possible, i.e., metadata.

- Use mathematical formalisms for the specification in order to avoid fuzziness and ambiguities.

Actually, the last two points are somewhat controversial. It is certainly good to have a great amount of metadata and rigid mathematical formalisms, but this needs also to be realized somehow. It has often been debated "how much metadata is necessary?" For example, the FGDC metadata content standard (see section 2.4.5 on page 28) defines a large and almost exhaustive set of data documentation attributes. How many of them need to be included in a dataset's documentation? There have been proposals for such "minimal metadata sets" with the motivation that the amount of metadata needs to be compatible with the *people* creating datasets, and eventually, with the data production cost. As soon as it is too complicated to create and maintain metadata, no one will supply it (Golder, 1990, p. 43).

Also the demand for rigid specifications can be dangerous with respect to its ability to be realized. The use of formal methods has often been advocated for the production of correct and reliable software in systems of ever increasing complexity (Bowen & Hinchey, 1995). For example, Frank and Kuhn (1995) propose OGIS specification with functional languages. Using formal methods allows one to be very precise about the system specified and to prove certain system properties, e.g., using mechanical proofing tools. Moreover, formal methods can also be used for the system *testing* by providing test patterns and

---

[6] Actually, low semantic accuracy sometimes might lead to many semantic heterogeneities which are *not detected*, exactly *because of* missing semantic information.

verifying test results[7]. The problems, however, with such formal methods are that they are difficult to understand and that there are myths (e.g., high costs) associated with projects based on formal methods. This means that people will usually refrain from using formal methods unless there are fervent formal method advocates involved. The degree of formalization needs, therefore, to be adapted to the skills of the personnel involved[8]. In a standardization process, it is somewhat difficult to determine a sensible formalization level since a standard is designed for as many uses as possible. Nonetheless, a rigid specification can provide semantic accuracy and it makes sense to use as much formalization as possible.

3.4 Open Geodata Interoperability Specification

3.4.1 Overview

The Open Geodata Interoperability Specification (OGIS) is an attempt to provide a specification for a computing environment providing geodata interoperability and addresses the impediments mentioned above. It is driven by the *OpenGIS consortium* which has most major GIS vendors as members. The overall objective is summarized by the OpenGIS consortium (Buehler & McKee, 1996, p. 8) as:

> "The Open Geodata Interoperability Specification provides a framework for software developers to create software that enables their users to access and process geographic data from a variety of sources across a generic computing interface within an open information technology foundation."

The specification process is still ongoing at the time of writing, but the overall framework has already been defined. This framework consists of three parts (Buehler & McKee, 1996, p. 4):

Open Geodata Model (OGM) Defined as "a common means for digitally representing the Earth[9] and earth phenomena, mathematically and conceptually" it provides a *canonical data model* necessary for integration.

OGIS Services Model A common specification model for implementing services for geodata access, management, manipulation, representation, and sharing between *Information Communities*.

---

[7] If functional languages are used such as in (Frank & Kuhn, 1995), then the formal specification is executable itself and allows a simple machine-based checking of the specification.

[8] "Thou shalt formalize but not overformalize" as Bowen and Hinchey put it (Bowen & Hinchey, 1995, p. 57).

[9] The reference to *earth* is given explicitly in the term *geodata*. OGM is tied to the Earth, i.e., it cannot be easily applied to other reference systems such as other planets or scale ranges (e.g., lunar mapping or molecular modeling).

Information Communities Model A framework for using the Open Geodata Model and the OGIS Services Model to solve not only the technical non-interoperability problem, but the institutional non-interoperability problem, such as various kinds of semantic heterogeneity (see also section 3.2.3).

The OGIS approach is a standardization – a *contract* between organizations – which goes beyond data format standardizations typically used for data migration. It adopts an object-oriented approach and offers a set of well-defined interfaces for writing software that interoperates with other OGIS-compliant software. The following sections will give a short overview of the Open Geodata Model and the OGIS Services Model.

3.4.2 Open Geodata Model

The Open Geodata Model defines a general and common set of basic geographic information types that can be used to model the geodata needs of more specific application domains, using object-based and/or conventional programming methods (Buehler & McKee, 1996, p. 11). The basic design guidelines for the geodata models were (Buehler & McKee, 1996, p. 15, emphasis added):

- "Independence of programming language, hardware, and network representations.

- Conformity to a *well-understood model* of earth features, in terms of their definition, spatial extent, properties, and relationship to other features.

- Supporting the universe of current, emerging, and future *modeling paradigms*, including concurrent support for feature- and layer (or coverage)-based paradigms, with mappings between them.

- Providing an *unambiguous definition* of basic geometric data types.

- Support of the *temporal dimension.*

- Support of *user extensions* to available data types, including possible dynamic (run-time) capability.

- Providing a well-defined interface supporting mechanism for the purpose of *geodata interchange.*

- Providing mechanisms for describing *spatial reference systems* that enable basic earth-centered transformation (i.e. projections) as well as space-based or arbitrary frames of reference.

- Providing mechanisms for describing *metadata* about data collections, including entity-attribute schemata.

- Providing a consistent, comprehensive set of geodata *behaviors/methods* expressible as defined types in conventional and object-based software.

- Harmonizing, to the extent possible, with *other geodata standards.*"

The specification of the OGM and other OGIS components is based on an object-oriented modeling approach proposed by Cook and Daniels (1994) and uses conventions for diagrams proposed therein. The primitives for these diagrams are shown in figure 3.1. This modeling approach proposes three levels of design (Cook & Daniels, 1994, p. 10):

Essential model This provides a description of real-world situations and is on the highest abstraction level.

Specification model Specification of *software* on a high level of abstraction.

Implementation model Translation of the specification model into a concrete environment, e.g., into a software system written in a particular programming language on a particular operating system and hardware platform.

OGIS itself is on the level of a specification model and is based on an essential model which is also part of the standardization. The Open Geodata Model identifies two fundamental geographic types which correspond to the object- and field-views mentioned before (Buehler & McKee, 1996, p. 40):

- A *feature* is "a representation of a real world entity or an abstraction of the real world." Features have a spatio-temporal *location* as attribute and are managed in *feature collections*. The location is given by a geometry and the corresponding spatio-temporal frame of reference.

- A *coverage* is "an association of points within a spatial/temporal domain to a value of a defined data type." The data type need not be a simple type (e.g., numeric type) but can be any compound, complex data type. Coverages represent what is discussed as *continuous fields* in chapter 4 and define basically a *relation* or *function* associating elements of a spatio-temporal definition domain $\mathbb{D}$ with elements of a value domain $\mathbb{V}$.

The essential model of a feature is shown in figure 3.2. A feature is basically a collection of identified properties. Each property (attribute) is related to a property value. For geometric properties, the property value is a coordinate geometry and is associated to a spatio-temporal reference system which can be used to derive the location of the feature. Each feature has a *unique identity* which is given by its *object identification (OID)*[10].

---

[10]At the time of this writing, it is discussed, but not yet clear, how the uniqueness of such an ID can be guaranteed in a distributed environment.

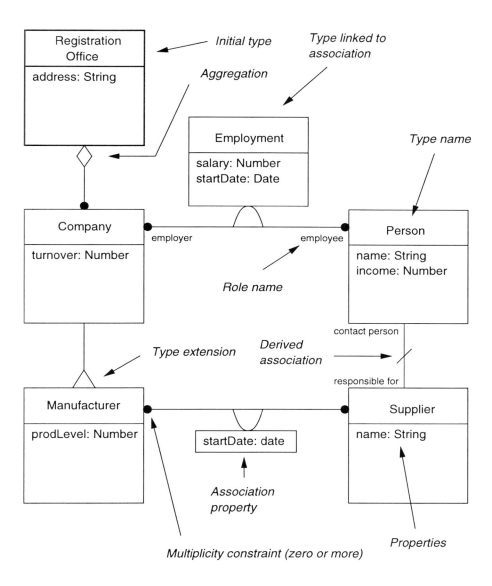

Figure 3.1: OGIS diagram conventions (Cook & Daniels, 1994, p. 361)

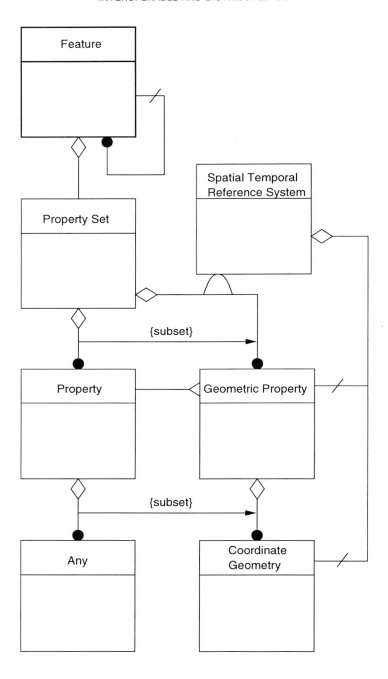

Figure 3.2: Essential model of OGIS feature type (Buehler & McKee, 1996, p. 42)

Using this generic feature model, a coverage can be seen as a *specialization* of a feature. The essential model in figure 3.3 shows that the geometric property defines the domain $\mathbb{D}$ of the coverage and that there is another property that defines a so-called *stored function* over the domain $\mathbb{D}$. This stored function provides the mapping of any point $s \in \mathbb{D}$ to a value $z(s)$ of the coverage. The stored function defines the *type* of its return value. Note that there are *no geometric details* about the inner structure of the coverage made explicit. That is, if the coverage is defined, for instance, by a regular grid, it is, in principle, not necessary to reveal information about the grid's geometry. However, the *specification models* for coverages define a set of specialized coverages such as grid coverages and point-set coverages (OGIS, 1996, p. 58).

Geometries in OGIS are based on so-called *well-known structures (WKS)* which are the basic building blocks for any geometry. A WKS can be constructed by a finite set of points which are called its *corners* (OGIS, 1996, p. 11). Every WKS can contribute to various features and the geometry of every feature can be defined by many WKS. There is a set of general types of WKS (so-called well-known types) defined in the specification model, covering types such as points, lines, surfaces etc.

3.4.3 Services Model

The OGIS Services Model defines the model for OGIS-compliant software components that can be used to build applications. Such a service is defined as "a computation performed by an entity on one side of an interface in response to a request made by an entity on the other side of the interface" (Buehler & McKee, 1996, p. 100). This means that services represent stand-alone functions which are not necessarily bound to any object (as object methods are) and provide a basic set of functionality (Buehler & McKee, 1996, p. 69):

- "The means by which Open Geodata Model data types can be collected to form complex models, queried for selections of subpart, and cataloged for sharing (both internal and external to Information Communities).

- Mechanisms for defining and creating Information Communities (...) and for developing linkages between these communities.

- The means by which Open Geodata Model data types, user-defined data types, and other capabilities (...) can be defined and their operations executed."

These services are grouped into categories addressing various aspects of an interoperable geoprocessing environment. As the OGIS standardization process is ongoing at the time of writing, the services specified are by no means meant to be complete and final. Most of these services correspond somehow to aspects of metadata in traditional data sharing and provide the

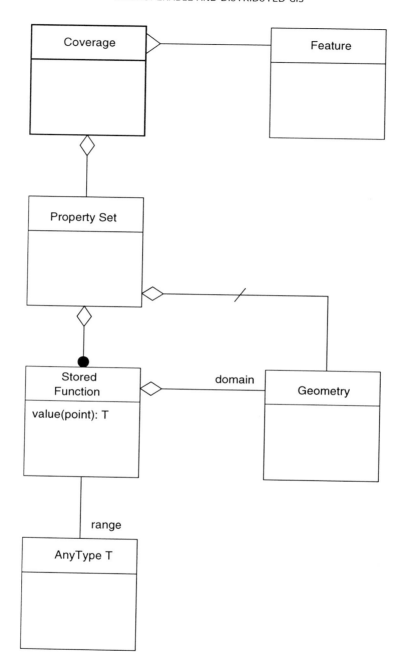

Figure 3.3: Essential model of OGIS coverage type (Buehler & McKee, 1996, p. 44)

mechanisms that an application can use to become informed about its context, e.g., which data are available, what the data mean, what can be done with them and so on. The service categories identified in OGIS are:

Feature registries Services that provide information about the *schema* of a feature, i.e., definition of the specific feature type. These schemata are managed by so-called *feature registries* which define the set of properties for every proper feature type and can be queried for that information. Moreover, feature registries provide *factories* which allow a feature of a given type to be instantiated.

Catalog registries Catalogs are directories of features and the primary access path for applications to the features in, for instance, a database. Each catalog indexes a set of features and other catalogs, i.e., a catalog can also contain other catalogs and as such represent a hierarchy. Catalogs define their own structure with corresponding *catalog schemata*, i.e., the structure of a catalog can differ from application to application and from information community to information community.

Spatio-temporal reference systems OGIS specifies a set of services to *register* various spatio-temporal reference systems and transformations between such reference systems. This is needed for various objectives, such as (Buehler & McKee, 1996, p. 74):

- "Users have a need to change spatio-temporal reference systems,
- The coordinate geometries connected with features are to be understood,
- Information is shared between information communities using differing spatio-temporal reference systems, or
- Implementations use differing parameter sets to implement spatio-temporal reference systems."

Semantic translation Services which provide a mechanism for the translation of features between information communities. These translators are meant to provide a mapping between corresponding representations of a feature in different information communities[11].

Operation registries Operation registries contain geoprocessing operations that can be invoked on geographic data. Such operations might be generic operations for the access of features in catalogs, geometric algorithms, spatial indexing techniques and so on. For that purpose OGIS

---

[11]It is, However, not very clear how such translation will happen. The OGIS guide (Buehler & McKee, 1996, p. 76) notes that "(...) unlike geodata modeling and distributed geodata access, automated semantic translation has not been part of the mainstream academic or commercial geoprocessing research agenda." Thus, it is expected that more research in this area can lead to better understanding of such semantic translation and a revision of the model.

defines "free" operations which need not necessarily be part of an object. Using such operations it is possible to provide, for example, an operation that performs a specific task for *multiple objects*. A typical example is an intersection operation which might be able to intersect various geometric primitives and which can be called by implementors of more complex, compound geometries.

Type registries Type registries provide the description of all types used in an OGIS environment. The type registry can be queried for existing types and be updated with new types. Moreover, similar to the feature registry it provides so-called *type factories* which can be used to instantiate objects of the respective type. Type registries can be understood as the major data dictionary of an OGIS environment.

Traders Traders are intermediate directories of "items of interest, such as catalogs and their contents" (Buehler & McKee, 1996, p. 86). In a traditional context, traders can be understood as meta-databases. However, the trader concept described in OGIS offers a much more versatile relationship between data producers and data users (or catalogs and catalog clients) in that a trader can actually "mediate" between them, and, for instance, assess fitness-of-use.[12]

Query services Queries are the services used to retrieve features from feature collections and catalogs. The reason to include queries as a special service category (and not as a "simple" operation on catalogs) is first, that queries can result in large collections. The management of such collections needs special mechanisms. For instance, it is not uncommon that the entire collection does not fit into primary memory, i.e., some kind of *database cursor* concept has to be applied. Second, queries might also have substantial computational complexity and are, therefore, candidates for asynchronous execution. Asynchronous operation in turn needs multiple stages such as initialization, start, suspend, stop, and status inquiry. This means, that the usual simple operations are not suitable to represent an entire query life cycle.

3.5 The Virtual Dataset concept

The approach taken here is similar to OGIS and is called the *Virtual Dataset (VDS)* (Stephan *et al.*, 1993; Bucher *et al.*, 1994; Včkovski, 1995; Včkovski & Bucher, 1996; Včkovski, 1996). The main idea is again that data exchange is not specified by a standardized data structure (e.g., a physical file format) but by a set of *interfaces*. These interfaces provide data access *methods* to retrieve the actual data values contained in a data set or a query result. This enables

---

[12]One can also envision a commercial usage in the real sense of "traders", e.g., in that traders provide the authorization and billing service for a dataset and its usage.

a data set to include "intelligent" methods to provide the data requested and helps to overcome the problems associated with the diversity of geographic information and semantic heterogeneities in general. Other approaches using a similar architecture are discussed in (De Lorenzi & Wolf, 1993; Voisard & Schweppe, 1994; Leclercq et al., 1996).

An application that uses a VDS does therefore not "read" the data from a physical file (or query a database), but it will call a set of corresponding methods defined in the VDS which return the data requested. Depending on the application domain, a VDS might just return pre-stored information, calculate the requested information or even provide "real-time" information by querying some measurement devices. The latter case is especially interesting in facility management applications where real-time aspects are very important. Consider for example a telecommunications network (e.g., telephone lines). If an application accesses the data (e.g., the network's geometry and topology) directly from a database, then the application needs to be able to perform all necessary operations to query the network's current status, i.e., it makes the application complex and fragile with respect to changes in the system. Embedding these mechanisms in "data" allows the application to reduce its responsibility and thus its complexity. It is the "job of the data set" to be able to provide the necessary data about itself. As soon as there is, for example, new equipment in the network which needs other techniques to query the operating status, this can be embedded as a new specialization in the dataset not requiring any change in the control application.

This concept is not new and is used in many different situations throughout the computer industry and research community. A typical example from a different (but nonetheless similar) area is the technique used to access various peripheral hardware components within a computer system. The discussion here will focus on printing devices, but it applies more or less to all other peripherals. In the late sixties and early seventies, computing happened to a large degree within products of the same manufacturer, i.e., proprietary systems were used, and there were almost no compatibility problems. The rise of the industry then yielded many hardware manufacturers who did not sell entire systems (i.e., everything from central processing unit to the appropriate furniture) but specialized products such as printers, disk systems, and so on. These products often had superior quality and lower prices than the products available from the all-in-one company. The customers, therefore, wanted to have the "freedom of choice" and be able to use whatever add-on products they wanted with their computer system.

For printing devices, this meant that people started to buy printers from companies that were not the original manufacturer of the main system. These printers often had characteristics slightly different (mostly improved) from their counterparts of the "original" supplier and the customer wanted to use these new features, such as printing boldface, changing line spacing and so on. The software which produced print output usually dealt with such printer control itself, i.e., it contained "hard-coded" control sequences to use the special printer features. It was soon realized that a more flexible scheme should be adopted because for every new printer type the software source code had to be changed and extended.

The next step was then to provide a static description of each printer's capabilities in a printer database and to use this information to control the printer. Such databases did contain, for example, control sequences[13] to eject a page from the printer, to switch the printer to boldface printing and so on. Therefore, an application only needed to know the format of the database and could subsequently use any printer that was defined in this database without hard-coding all the printer's special features. This technique was still in use until the late eighties, even in mainstream text processing applications on personal computers.

The advance of printer technology and the growing number of competing vendors then led to other approaches to support device-independent printing for various reasons, in particular, because of the lack of maintainability and flexibility. The technique of parameterized description as mentioned above was in general very difficult to maintain. On the one hand, parameters for a large variety of printer hardware had to be supplied. The software vendor usually did supply a set of parameters for the more frequently used printer models and also offered methods to update the database with new parameter sets. On the other hand, new capabilities forced changes in the parameter scheme of a specific application. A new printer might have had a new key selling feature which allowed superscript- and subscript-printing. A competitive software product supporting this new printer model needed to include the printer's new features, therefore requiring a parameter scheme change in the printer database to support sub- and superscripting. Some systems used *flexible* parameter schemata which were described, for example, by a kind of *printer definition language*, much like the data definition languages discussed in section 3.3.1. This approach did enhance flexibility, but it also increased complexity on the application side. An application needs to understand the printer definition in order to correctly address the printer.

The solution to these maintenance problems clearly needed a more flexible and versatile way of shielding the device-dependency from the application. The approach taken by most software systems today is based on individual software modules called *drivers*. These drivers are device-specific programs with well-defined interfaces towards the calling application or the operating system, respectively. An interface definition is analogous to the definition of a parameter scheme. The implementation of the interface however offers much more flexibility, since it leaves the details for, say, selecting a specific printing typeface, within the driver. This means also that the application (or operating system) does not *assume* something about the technique used to control printers, e.g., the assumption that printers are controlled by sending a sequence of characters (identified by a special prefix-character such as ASCII-

---

[13]The sequences are often called *escape sequences* due to code-name of the prefix character.

Escape) embedded in the real "data". This flexibility was needed, for example, to support devices which used totally different techniques for the transfer of page content such as POSTSCRIPT-printers (see also section 6.2 of chapter 6).

In parallel with this development of hardware addressing from direct manipulation over parameterized descriptions to procedural descriptions (drivers), there was a general movement towards layered software architecture which led to the notion of *operating systems*, i.e., software systems that shield user applications from hardware details and offer a set of common functionality. This means that the parameter scheme or driver interfaces, respectively, were defined mostly by the operating system vendor and not by the end-user application vendor. In the past decade, user applications hardly ever communicated directly with drivers (or even used parameter sets) but only by using operating system services. It needs to be noted, however, that the amount of shielding typically varies greatly with the *type* of hardware addressed. Mass storage devices such as hard disks are hardly ever controlled directly by an end-user application, and the corresponding "driver" in most operating systems actually consists of a whole set of drivers, managing high-level components such as *file-systems* and low-level components such as *SCSI communication*. Specialized and infrequently used hardware, such as digitizers or tape robots, is often still addressed directly by end-user applications since operating systems do not offer corresponding abstraction layers.

The objective of the previous simplified discussion of the evolution of addressing of peripheral devices in computer systems was to show that a *layered architecture* with well-defined interfaces can reduce the complexity of a system while improving the versatility. The concept of the VDS is basically an adoption of this principle to the representation of spatial data. In terms of the *Open Systems Interconnection (OSI)* model of the International Standards Organization (Stevens, 1990) a VDS adds a new layer within the OSI application layer (layer 7). An application accesses data through a well-defined interface to the VDS. This technique is equivalent to software components called *middle-ware* which evolved in the last decade and became popular for client/server environments and especially for vendor-independent access to database management systems (DBMS), such as *Open Database Connectivity* (ODBC) (ODBC, 1992). The approach taken by VDS goes one step beyond (or above) platform- and vendor-independent data access and enters on a higher abstraction level, providing an interface to the data themselves and not their storage (as for example a DBMS). Referring to the discussion of peripheral devices, a VDS can be seen as a "driver" for a particular dataset. A VDS therefore is a *software component* and the range of possible implementations is bound only by the well-defined interface the VDS has to provide to the outer world. There are many examples where this flexibility offers advantages in the system design, such as:

Dynamic content In a real-time environment a VDS can provide current measurement values, i.e., provide data which are not stored in some database but retrieved directly from a measurement device. This is often needed in AM/FM-type applications such as network management or flight-control systems.

Multiple representations A VDS can provide a context-specific representation of its content, where the context is given, for example, by the visualization scale and desired level of detail.

Derived values Derived values can be easily defined by corresponding methods of the VDS.

Simulations A VDS can deliver *simulated* values for various kinds of applications. The values could be calculated on the base of underlying measurements (e.g., stochastic simulation), or based on a (parameterized) model for applications such as controlled system testing.

A simple yet useful application of VDS is the *simulation of continuity* by interpolation which is used as an example in this thesis. In particular, the representation of *continuous fields* is a useful example for a VDS. The underlying essential model for continuous fields is therefore discussed in a separate chapter (chapter 4).

Flexible uncertainty representations are another useful example of a VDS's object-oriented features. Uncertainty modeling techniques are needed on the one hand for a reliable representation of continuous fields. On the other hand, uncertainty models are a *fundamental*, yet often, neglected component of a scientific computing system, and therefore, discussed in a separate chapter as well (chapter 5).

The architecture proposed is a three-tier model (see figure 3.4), where the middle tier represents the *virtualization layer* provided by VDS. The VDS objects encapsulate "business logic" and serve as an intermediary between data user and static data. The domain specific knowledge to define the methods of a VDS is available in the data *producer's* domain. This means that there is a shift of responsibility from data user to data producer. A data producer ideally delivers a dataset as a VDS, i.e., as an object embedding the methods necessary to allow a data user to query a VDS for the data needed and a VDS to answer such queries, respectively. The details of the "data" part of the architecture are left open to the VDS implementation. Data sources could be:

- remote or local database management systems,

- files in a local or remote file-system,

- objects accessed by *uniform resource locators (URL)*, see section 6.3.4 in chapter 6,

- remote or local measurement devices,

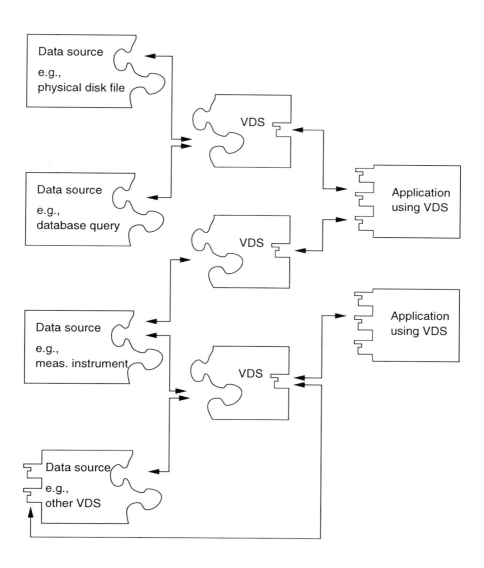

Figure 3.4: Three-tier architecture of Virtual Data Set

- computation servers
- Virtual Data Sets[14] or any other data sources.

The point is that a data user is not aware at all of *where* and *how* a VDS obtains its source data, or *state* in object-oriented parlance. Actually, a data user *need not* be aware of data sources, but instead can query a VDS for this information as part of descriptive information ("metadata") that a VDS has to supply. The embedding of all that intelligence into a VDS seems to heavily increase a dataset's complexity. Consider for example a VDS having a remote measurement protocol as data source. The VDS needs to be able to handle the *entire communication* with the measurement device, i.e., the dataset will be a complicated piece of code compared with an alternative of transferring the measurement values as simple numbers in a text-file. Yes! But, eventually, someone *has* to handle that communication. The creation of VDS should follow the same principles used elsewhere in software engineering, e.g., modularity and software reuse. It makes sense, therefore, to embed the communication with the measurement device, e.g., in a (class) library which can be used by a VDS. Another example are spatial interpolation methods that are embedded in a VDS in order to provide values at unsampled locations. It makes no sense, of course, to have a programmer write spatial interpolation methods from scratch for every VDS. A set of predefined methods from a library or an OGIS service (see section 3.4.3) could be used and specialized within a VDS. In that sense, VDS can be seen as *generic* software components. An implementation should therefore be based on one of the emerging standards for software components. Chapter 7 discusses several alternatives based on current standards for software components.

3.6 Review

The approaches presented in this chapter address interoperability by defining a set of common interfaces. These interfaces provide a well-defined access to geodata and geoprocessing services for interoperating applications. The geoprocessing envisioned is a distributed computing environment based on many small software components which interoperate on the basis of these interfaces. The components represent all functional elements needed in the overall system, including datasets (as VDS), user interfaces and the like. The main advantage of using interfaces instead of data formats is the increased flexibility. Users of software components do not need to be aware of what happens "behind the scenes" as long as the components comply with the agreed interfaces. Components which are data sources are not limited to access of persistent information but can also derive or acquire data dynamically.

The success of such approaches depends on the specification of the interfaces. Are the interfaces general enough to allow many different types of

---

[14]This allows *piping* similar to the Unix computing model (Stevens, 1990, p. 102–109).

components or VDS? Are the interfaces specific enough to avoid the necessity of post-processing the data retrieved? Such questions need to be answered by corresponding essential models which provide the necessary understanding of the real-world entities represented by an interface. The next part provides essential models for continuous fields and uncertainty representations. These models are meant to be an example of questions that need to be addressed when defining essential models.

II

Essential models

CHAPTER FOUR

Example: field representations

4.1 Overview

Data describing *continuous fields* are a major information type used in many natural sciences and spatial data processing applications. This chapter discusses the characteristics of continuous fields as they are used in data processing systems. The overall objective is to show that an approach like VDS can help to overcome the impediments resulting from the necessary discretization when sampling continuous fields. A general (essential) field model is developed which can be used as a theoretical basis for the implementation of corresponding VDS and for the identification of individual components and properties.

The first part of this chapter defines what is understood as (non-random) continuous field and discusses the sampling of such fields on a theoretical basis. This is followed by a discussion of current representation techniques and their advantages and disadvantages. Analyzing the impediments of using such static representations finally leads to the specification of the requirements for an implementation based on the VDS concept.

4.2 Continuous fields

The concept of a *field* in physics denotes a phenomenon whose quantity depends on the location in a space. The corresponding mathematical model therefore describes the quantity as a *function* $z(\cdot)$ of elements s (locations) of that space $\mathbb{A}$. Typically, the field is defined on a subset $\mathbb{D} \subset \mathbb{A}$ which is called *(definition) domain*[1] of $z(\cdot)$. The values of $z(\cdot)$ describe the physical

---
[1] $\mathbb{D}$ is sometimes also called *support*. More rigidly defined, the support of a function $f(\cdot)$ is the set A for which $f(x \in A) \neq 0$.

quantity and we will denote their *range* or *value domain* as $\mathbb{V}$, i.e., $z(s) \in \mathbb{V}$. The field can therefore be written as[2]:

$$z : s \longrightarrow z(s), \quad s \in \mathbb{D} \subset \mathbb{A}, \quad z(s) \in \mathbb{V} \qquad (4.1)$$

We will assume that $z(\cdot)$ is a *measurable function*, i.e., that the domain $\mathbb{D}$ is a *measurable set* and that a *measure* μ is defined on the space $\mathbb{A}$. This assumption is not limiting since $z(\cdot)$ represents a physical quantity and the possible values are bounded due to physical reasons (i.e., if $\mathbb{V} \subset \mathbb{R}$ there are two numbers $a, b \in \mathbb{R}$ with $a < x$ and $b > x$ for all $x \in \mathbb{V}$).

We will further assume that the space $\mathbb{A}$ is an *infinite* and dense set. We call a field *continuous* if the domain $\mathbb{D} \subset \mathbb{A}$ is also infinite. It is important to note the fundamental difference between the notion of *continuous function* and *continuous field*. The first relates to continuity in the *range* $\mathbb{V}$ whereas the latter refers to the *continuity* in the *domain* $\mathbb{D}$. A continuous function is a function for which a small change in the independent variable causes only a small change in the dependent variable (the function value). In contrast to this, the term *continuous field* means that the domain $\mathbb{D}$ of the function $z(\cdot)$ is a *dense set*. This means that – assuming $\mathbb{A}$ is a topological space – each neighborhood of a point (element) $x \in \mathbb{D}$ contains also another point $y \in \mathbb{D}$ with $x \neq y$.

For this discussion we will further assume that the range $\mathbb{V}$ is a subset of $\mathbb{R}^n$. If $n = dim \mathbb{R}^n = 1$ then the field is called a *scalar field*, otherwise it is a *vector field*[3]. The space $\mathbb{A}$ is almost always a metric space (i.e., a space with a defined metric $\delta(\cdot, \cdot)$) and equal to $\mathbb{R}^m$ with $m = \{1, \ldots, 4\}$.

Table 4.1 shows some examples of fields and their domain and range. It has to be noted that $\mathbb{A}$ or $\mathbb{D}$, respectively, are sometimes inhomogeneous in the sense that they consist of "subspaces"[4] that have different physical meaning. Consider, for example, a field on a domain which is a subset of physical space and time ($\mathbb{M}$). In that case, one might set $\mathbb{A} = \mathbb{M} = \mathbb{R}^4$ with three spatial coordinates and one temporal coordinate. Unless there is a specific metric defined (e.g., Minkowski-metric), $\mathbb{M}$ must actually be seen as a Cartesian product $\mathbb{S} \times \mathbb{T} = \mathbb{R}^3 \times \mathbb{R}$. A field $z(\cdot)$ on the four-dimensional space–time $\mathbb{M}$ can therefore be seen as a function $f(\cdot)$ on three-dimensional physical space $\mathbb{S}$ mapping every point $s \in \mathbb{S}$ to a function $g(\cdot)$ of the time axis $\mathbb{T}$ ($\mathcal{C}(\mathbb{T})$ is the set of all functions on $\mathbb{T}$):

$$\begin{aligned} z &: \mathbb{M} \longrightarrow \mathbb{V} \\ f &: \mathbb{S} \longrightarrow \mathcal{C}(\mathbb{T}) \\ g &: \mathbb{T} \longrightarrow \mathbb{V}, \quad g \in \mathcal{C}(\mathbb{T}) \end{aligned} \qquad (4.2)$$

This means that a value $z(s)$ is given by

$$z(s) \;=\; [f(s|_\mathbb{S})](s|_\mathbb{T})$$

---
[2] We will make no distinction between the *physical* field and its *mathematical* description.
[3] I.e., a scalar field is a special case of a vector field.
[4] The quotes are used because these are not actually proper subspaces.

$$= g(s|_{\mathbb{T}}) \quad \text{with} \quad g = f(s|_{\mathbb{S}}) \tag{4.3}$$

The following discussion of the measurement of fields nonetheless assumes that $\mathbb{A}$ is homogeneous and silently expects a splitting into homogeneous subspaces as given by (4.2) if $\mathbb{A}$ is not homogeneous.

4.3 Measurement of fields

The overall objective of this chapter is to show how a VDS can improve the digital representation of continuous fields and to provide a corresponding essential model. Thus, it is necessary to analyze and understand the relationship between the *field* and the *field samples*, i.e., the measurement process when sampling a field.

Every measurement of a physical quantity needs a *measurement model*. This model basically defines a mapping between the possible states of the quantity into (usually) rational numbers, i.e., a mapping from "reality" into a mathematical object. This is one of the most important and useful epistemic principles in natural science. Relating real world objects and their states to mathematical objects enables the use of mathematical rules applied to the mathematical objects to derive insights for the real world objects. This mapping – the measurement model – is an idealization and never bijective. This can be due to inherent variability in the real world object (i.e., states cannot be identified exactly), measurement errors, and so on. To account for this idealization, *uncertainty models* such as probabilistic approaches or fuzzy techniques are used (see chapter 5). In this section, we will assume that the measurement model is already included in the *mathematical* description of a field $z(\cdot)$ and discuss only the measurement (or selection) of individual values from $z(\cdot)$. The previous definition of the term *continuous field* identified $\mathbb{D}$ as an *infinite* set. The function $z(\cdot)$ relates *infinitely many* points $s \in \mathbb{D}$ to values $z(s) \in \mathbb{V}$. The measurement of a field, i.e., the description of the field with a *finite* set of numbers, will inevitably reduce the *information content*, or Shannon-entropy (see page 99), of the field. This is the main cause of the impediments when using data representing fields discussed later on. Here, we will investigate in more detail what the measurement of a field involves.

Let $z(\cdot)$ be a continuous function on $\mathbb{D}$, i.e., an element of the space $\mathcal{C}(\mathbb{D})$ of all continuous functions on $\mathbb{D}$. The range $\mathbb{V}$ shall be one-dimensional and equal to $\mathbb{R}$ for the sake of simplicity[5]. The measurement of the field results in a set of numbers $\{z_1, \ldots, z_n\}$ which are somehow related to the field $z(\cdot)$. The relation f_i between a z_i and $z(\cdot)$ can be characterized as follows:

1. f_i is a *functional*, a mapping from $\mathcal{C}(\mathbb{D})$ into $\mathbb{V} = \mathbb{R}$.

2. The value z_i is given by $z_i = f_i[z(\cdot)]$.

---

[5]The extension to non-scalar fields is straightforward, given an orthogonal decomposition of a multidimensional $\mathbb{V}$ into one-dimensional subspaces $\mathbb{V}_i$.

Field	dim($\mathbb{D}$)	dim($\mathbb{V}$)	Indep. variables ($\mathbb{D}$)	Dep. variable ($\mathbb{V}$)
$T(z)$	1	1	z: Spatial coordinate (height)	T: Temperature of the Atmosphere at height z
$E(t)$	1	3	t: Time coordinate	E: Electrostatic force as time-series
$H(x,y)$	2	1	x,y: Spatial coordinates	H: Height of orography (elevation)
$P(x,y,z)$	3	1	x,y,z: Spatial coordinates	P: Soil porosity
$v(\lambda,\phi,z)$	3	3	λ,ϕ,z: Spatial coordinates	v: Wind vector
$\sigma(x,y,z)$	3	9	x,y,z: Spatial coordinates	σ: Stress tensor
$\Theta(\lambda,\phi,p,t)$	4	1	λ,ϕ coordinates (longitude, latitude), p: pressure level, t: time	Θ: Potential temperature at (λ,ϕ,p,t)
$\Theta_t(\lambda,\phi,p)$	3	∞	λ,ϕ coordinates (longitude, latitude), p: pressure level	Θ_t: Time-series of potential temperature at (λ,ϕ,p)
$I(x,y,z,t,\lambda)$	5	1	x,y,z: Coordinates of a point in atmosphere, t: time, λ: wavelength	I: Intensity of radiation at x,y,z at time t of wavelength λ

Table 4.1: Some examples of fields

3. It makes sense to assume some regularity for f_i. For example, linearity of f_i is often desired, i.e., f_i is a linear functional fulfilling

$$f_i[\alpha_1 z_1(\cdot) + \alpha_2 z_2(\cdot)] = \alpha_1 f_i[z_1(\cdot)] + \alpha_2 f_i[z_2(\cdot)] \qquad (4.4)$$

In many cases z_i is the value of $z(\cdot)$ "at" location s_i. If s_i is a point (i.e., not a *set* of points) z_i is written as $z_i = z(s_i)$. However, a measurement apparatus is never so accurate that it can sample *exactly* the field's value at location s_i. The sampled value z_i is rather a kind of weighted average of $z(\cdot)$ centered around s_i. We can therefore write the assignment, i.e., the functional f_i, as:

$$z_i = \int_{\mathbb{D}} z(s)\phi_i(s)ds \qquad (4.5)$$

with f_i given by

$$f_i(\cdot) = \int_{\mathbb{D}} \cdot \phi_i(s)ds \qquad (4.6)$$

The functional f_i is thus related to a function $\phi_i \in \mathcal{C}(\mathbb{D})$. ϕ_i will be called the *selection function*[6]. Using the usual definition of scalar product $\langle \cdot, \cdot \rangle$ in $\mathcal{C}(\mathbb{D})$, we can therefore write

$$z_i = \langle z, \phi_i \rangle \qquad (4.7)$$

The selection function ϕ_i depends on all stages of the measurement process and has ideally a small support around s_i and its maximum over s_i. If f_i exactly retrieves the value at a point location s_i, then ϕ_i is the Dirac-"function" δ_{s_i} centered at s_i:

$$\begin{aligned} z_i &= \int_{\mathbb{D}} z(s)\delta_{s_i}(s)ds \\ &= \int_{\mathbb{D}} z(s)\delta(\|s - s_i\|)ds \end{aligned} \qquad (4.8)$$

The concept of the selection functions can also be used if s_i is a subset of $\mathbb{D}$ describing a region of $\mathbb{D}$. The following example shows ϕ_i for a measurement of a temporal average of the field $z(\cdot)$ with $\mathbb{D} = \mathbb{T}$, i.e., a time-series, in the interval $s_i = \{t | \tau_a \leq t \leq \tau_b\}$:

$$\phi_i(t) = \begin{cases} \frac{1}{\tau_b - \tau_a} & : \tau_a \leq t \leq \tau_b \\ 0 & : \text{otherwise} \end{cases} \qquad (4.9)$$

z_i is then given as:

$$\begin{aligned} z_i &= \int_{\mathbb{T}} z(t)\phi_i(t)dt \\ &= \frac{1}{\tau_b - \tau_a} \int_{\mathbb{T}} z(t)dt \end{aligned} \qquad (4.10)$$

---

[6] ϕ_i is sometimes also called *test function*, e.g., in the theory of distributions (Constantinescu, 1974, p. 31), or *kernel* in integral transforms.

For another example, consider the sampling procedure used in raster images. The measurement of the value at pixel i yields a sensor-specific average of $z(\cdot)$ within the location s_i. s_i is related to the sensor geometry (layout of the pixels) and the position of the sensor in space. The selection function ϕ_i can therefore be modeled similarly to (4.9), incorporating the sensor's characteristics. Figures 4.1 and 4.2 show some examples of selection functions.

However, there are cases when f_i cannot be written using a selection function as in 4.6. For example, if the measurement of $z(\cdot)$ "at" s_i yields the maximum value of $z(\cdot)$ in s_i:

$$z_i = \max_{s \in s_i} z(s) \qquad (4.11)$$

Here, there is no function ϕ_i satisfying form (4.6) for f_i. Rigidly, the theory of Schwarz's distributions or generalized functions needs to be applied here (Constantinescu, 1974).

The notion of $z(\cdot)$ as a function mapping $s \in \mathbb{D}$ to exactly one value[7] $z(s) \in Domain$ is not a realistic model for many natural phenomena because it neglects any randomness or uncertainty. These random variations may have differing causes, including inherent natural variation of the phenomena, all types of measurement uncertainties, and ignorance and complexity of the underlying processes. There are several ways to account for such uncertainties. For the sake of brevity, only probabilistic approaches are presented here, i.e., the modeling of each field value as a *random variable*. The modeling of uncertainties is discussed in a more general context in chapter 5 on page 91.

The relation in equation 4.1 is therefore extended to map every $s \in \mathbb{V}$ to a random variable $Z(s)$ ($Z(s)$ is sometimes called a *random function* (Isaaks & Srivastava, 1989; Deutsch & Journel, 1992)). The random variables for different s are not independent, so that we are actually considering a probability space over $\mathbb{V} \times \mathbb{D}$. For the subsequent discussion, we will use the *probability distribution function (cdf)* $F(z; s)$ to characterize the field as defined by (for a scalar field $z(s) \in \mathbb{R}$):

$$F(z; s) = P\left(Z(s) < z\right) \qquad (4.12)$$

The cdf fully characterizes $Z(s)$. Thus, the measurement of the random field also needs to account for the cdf. Often, it is suitable to model $F(z)$ by a *small* set of numbers. For example, the expectation value $E\left[Z(s)\right]$ and variance $V\left[Z(s)\right]$, given by[8]

$$E\left[Z(s)\right] = \alpha_1 = \int_{\mathbb{V}} z dF(z) \qquad (4.13)$$

---

[7]The value might also be a multidimensional quantity, i.e., a vector.

[8]α_i denotes the i-th moment about the origin with $\alpha_i = \mathcal{E}\left[X^i\right]$. β_i is the j-th central moment or moment about the mean (Daintith & Nelson, 1991, p. 220) defined by $\alpha_i = \mathcal{E}\left[\alpha_1 - X\right]^i$.

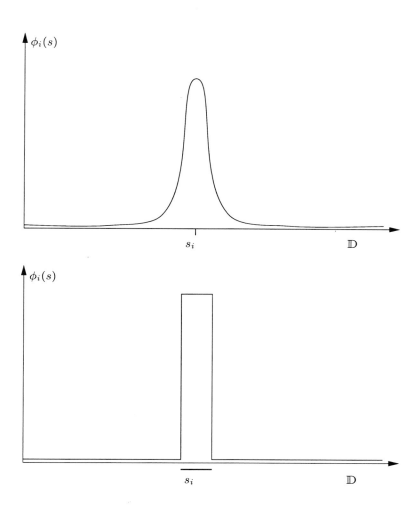

Figure 4.1: Examples of selection functions with $\dim \mathbb{D} = 1$

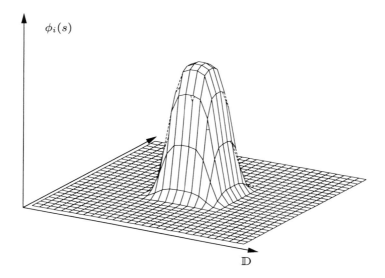

Figure 4.2: Example of a selection function with $\dim \mathbb{D} = 2$

$$V[Z(s)] = \beta_2 = \int_V (z - \alpha_1)^2 dF(z) \tag{4.14}$$

are often used to describe $F(z)$. There are many well established methods to estimate the moments α_1 and β_2 with descriptive measures such as *mean value* and *standard deviation*.

This example shows that for random fields we need *a set of numbers* to represent the field "value" at a location s_i, i.e., $\{z_{i,1}, \ldots, z_{i,m}\}$. The number of parameters used to describe the field value is usually small (e.g., $m = 2$ and $z_{i,1} = \alpha_1$ and $z_{i,2} = \beta_2$ as in the example above). There are cases, however, where the sampling of the phenomenon provides enough information to describe $Z(s)$ in more detail. An example is the characterization of $Z(s)$ using its empirical distribution function (histogram) with c classes:

$$\begin{aligned} p_j &= P(a_j < Z(s) \leq a_{j+1}) = F(a_{j+1}) - F(a_j) \\ j &= 1 \ldots c \\ a_1 &= -\infty, \quad a_{c+1} = \infty \end{aligned} \tag{4.15}$$

A field value at s_i is therefore represented by $2c - 1$ values:

$$\begin{aligned} z_{i,1} &= p_1 \\ &\vdots \\ z_{i,c} &= p_c \\ z_{i,c+1} &= a_2 \\ &\vdots \\ z_{i,2c-1} &= a_c \end{aligned} \tag{4.16}$$

Whereas in the case of non-random fields there is a single functional f_i (with an associated selection function ϕ_i) per value which maps the field $z(\cdot)$ to the sample z_i, random fields need a set of functionals $f_{i,j}$ (each possibly related to selection functions $\phi_{i,j}$) with

$$f_{i,j} : Z(\cdot) \xrightarrow{f_{i,j}} z_{i,j} \tag{4.17}$$

It is common that in a measurement for many "locations" i and uncertainty measures j, the selection functions remain the same for both the "spatial" part and the uncertainty part, i.e., the selection function $\phi_{i,j}$ can be separated into $\phi_{i,j} = \phi_i \circ \psi_j$. Moreover, the ϕ_i can be defined as a function of the location s_i, i.e., $\phi_i(\cdot) = \phi(s_i; \cdot)$. In the ideal case of a Dirac-δ-distribution, this would be:

$$\begin{aligned} \phi_i(s) &= \phi(s_i; s) \\ &\hat{=} \delta(s - s_i) \end{aligned} \tag{4.18}$$

The functions ψ_j represent the uncertainty measures and are, for example, given by the functionals that yield statistical moments (see eq. (4.13) and (4.14)).

If the functionals $f_{i,j}$ can be related to a spatial location s_i and an uncertainty measure type j, we can write a data set as a set of tuples:

$$(s_i; z_{i,1}, \ldots, z_{i,m}) \quad \forall i = 1, \ldots, n \tag{4.19}$$

These tuples define a relation between s_i and $(z_{i,1}, \ldots, z_{i,m})$, i.e., using the language of relational databases, the s_i can be seen as primary keys for the relation given by (4.19).

(4.19) is important in so far as it shows that the relation (4.1) mapping $s \in \mathbb{D}$ to $z(s) \in \mathbb{V}$ is reduced by the sampling to a mapping between a *finite* set of locations $\{s_i\}$ and tuples $(z_{i,1}, \ldots, z_{i,m})$. This mapping is defined by the functional f_{ij} or its decomposition into ϕ_i and ψ_j, respectively. This process is shown in figures 4.3 and 4.4. Extending (4.19) with a set M of additional dataset-specific information (*metadata*, see chapter 2, section 2.4.5) we can specify a formal notation of a generic dataset $\mathcal{D}$ representing a continuous field:

$$\mathcal{D} = \{M, \{s_i; z_{i,1}, \ldots, z_{i,m}\}_1^n\}, \quad s_i \subset \mathbb{D}, \quad z_{i,j} \in \mathbb{R} \tag{4.20}$$

4.4 Current representation techniques

This section discusses some of the techniques currently used for the representation of a continuous field. The various approaches are special cases of (4.20) and differ mainly in restrictions regarding the locations s_i. Recall that a "location" s_i is meant here to be a generalization in the sense that it represents a subset of the definition domain of the field. It might be a point-location in the four-dimensional space–time $\mathbb{M}$, a 2-dimensional region on the earth's surface, and so on.

In the context of spatial information systems, such representation techniques are called *spatial data models* (Goodchild, 1992) and are not to be confused with *spatial data structures*, i.e., issues of mapping a formal description into numbers that can be stored into a computer.

Goodchild (1992) identifies six different spatial data models (following the description of (Kemp, 1993)):

- Cell grids

- Polyhedral tessellation (polygons)

- Simplicial complexes (triangulated irregular networks)

- Lattice or point grids

- Irregular points

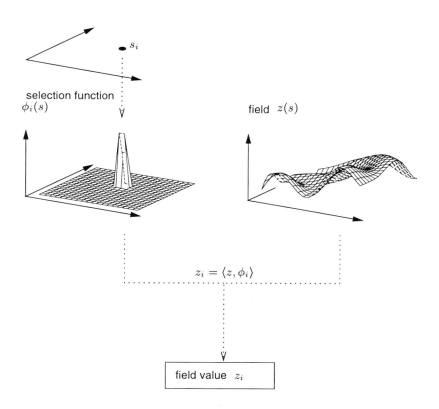

Figure 4.3: Discretization of continuous fields, s_i is a point

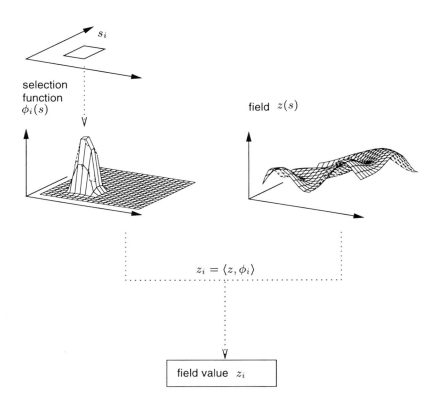

Figure 4.4: Discretization of continuous fields, s_i is a "pixel"

- Contour models

These various spatial data models vary in the choice of the spatial structure of the indices s_i and (sometimes) in the interpretation of the related field measurements $\{z_{i,1}, \ldots, z_{i,m}\}$. These six models are briefly described in the following, adding some extensions for dim $\mathbb{V} > 2$.

4.4.1 Cell grids

A cell grid is a tessellation of the study area $\mathbb{V}$ into cells s_i of the same size. The boundary planes of the cells are perpendicular or parallel to each other, i.e., the cells are intersections of $2n$ half-spaces $H_{k,l}$ ($n = dim\mathbb{V}$ and $k = a, b$, $l = 1, \ldots, n$). The half-spaces $H_{k,l}$ are defined by corresponding hyperplanes $h_{k,l}$ which obey:

$$\begin{aligned}(i) \quad & h_{k,l} \perp h_{k,m} \quad l \neq m \\ (ii) \quad & h_{a,l} \parallel h_{b,l}\end{aligned} \quad (4.21)$$

The cell s_i is defined by

$$s_i = \bigcap_{k=a}^{b} \bigcap_{l=1}^{n} H_{k,l} \quad (4.22)$$

with the half-spaces $H_{k,l}$ given by:

$$\begin{aligned}H_{a,l} &= \{x \in \mathbb{V} | x \leq h_{a,l}\} \\ H_{b,l} &= \{x \in \mathbb{V} | x \geq h_{b,l}\}\end{aligned} \quad (4.23)$$

Note that the "less than" and "greater than" relations in 4.23 refer to the position of x regarding the hyperplane $h_{k,l}$, i.e., if $x \leq h_{k,l}$ and $y \geq h_{k,l}$ then x and y are not on the same side of $h_{k,l}$. The relation is equivalent to comparing the sign of the scalar product of x and the normal of $h_{k,l}$. For $n = 2$ the s_i are rectangles; they are cuboids for $n = 3$. An equivalent formulation is the definition of a cell by the (convex) hull of 2^n vertices from a lattice (see section 4.4.4). Figure 4.5 shows an example of a two-dimensional cell grid.

4.4.2 Polyhedral tessellation

Polyhedral tessellation is a tessellation of the study area into polyhedra s_i of dimension $dim\,\mathbb{D}$, i.e., the polyhedra have the same dimension as the $\mathbb{D}$. This is a generalization of the cell grids mentioned above in the sense that the s_i are generally given as the hull of a set of m points, where $m > dim\,\mathbb{D}$. The polyhedra need not be convex. If $dim\,\mathbb{D} = 2$, this is sometimes called *polygon coverage*. In general it is assumed that two polyhedra s_i and s_j of the tessellation meet, if at all, in a common face or edge, respectively. Figure 4.6 shows an example of a two-dimensional polyhedral tesselation.

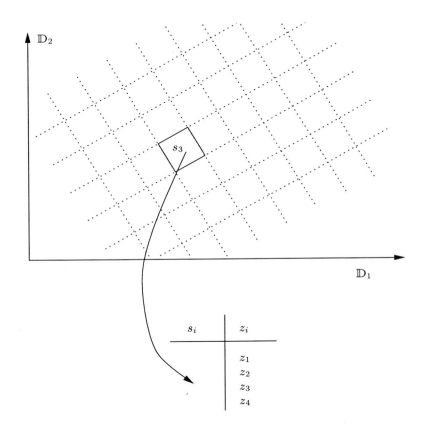

Figure 4.5: Cell grid with $\dim \mathbb{D} = 2$

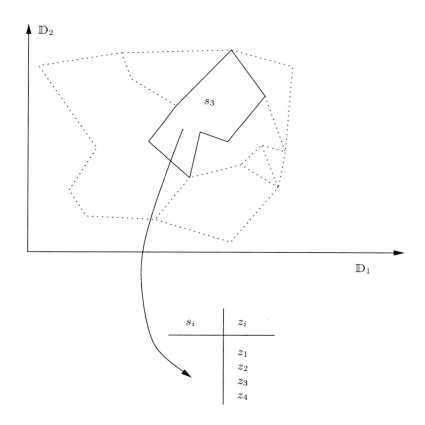

Figure 4.6: Polyhedral tesselation with $\dim \mathbb{D} = 2$

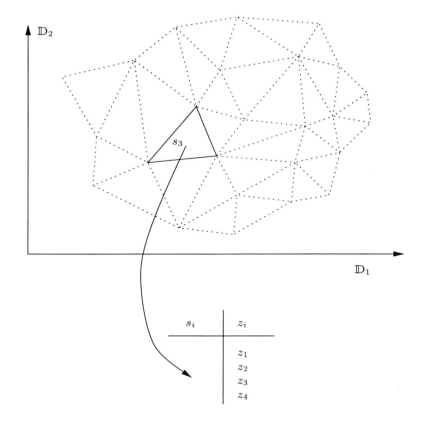

Figure 4.7: Simplicial complex with $\dim \mathbb{D} = 2$

4.4.3 Simplicial complexes

This is a special case of the polyhedral tessellation with $m = \dim \mathbb{D} + 1$. Each s_i is an m-simplex, i.e., the convex hull of m points or the intersection of m half-spaces. In the case of $\dim \mathbb{D} = 2$, the simplices are triangles and a simplicial complex is often called a *triangulated irregular network (TIN)*. The set of the simplices s_i sometimes obey additional conditions such as the *Delaunay-criterion* (O'Rourke, 1994, p. 175). Figure 4.7 shows an example of a simplicial complex. Note that the *entire* simplex is related to a field value and not the vertices of the simplex as it is usually identified in a TIN.

4.4.4 Lattice or point grids

The indices s_i are here the subset of a *lattice* with generator vectors $v_1, \ldots, v_m$. The indices therefore contain exactly one point of $\mathbb{V}$. However, this structure is equivalent (homeomorphic) to the cell grid since each point of the lattice

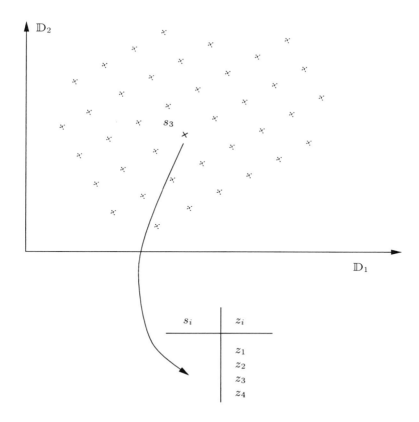

Figure 4.8: Point grid with $\dim \mathbb{D} = 2$

can be mapped to a corresponding cell[9] given by the hull of the 2^m points $\{s_i + \sum_{k=1}^{m} a_k v_k\}$ $\forall a_k \in \{0, 1\}$. In most cases, the generator is orthogonal (i.e., $\langle v_k, v_l \rangle = \delta_{kl}$). Figure 4.8 shows an example of a two-dimensional cell grid.

4.4.5 Irregular points

This is a generalization of the lattice. The s_i are points of $\mathbb{V}$, without implicit order. A homeomorphism can be defined here for polyhedral tessellation and for simplicial complexes. The mapping to a polyhedral tessellation is given by the corresponding Voronoi diagram (O'Rourke, 1994, p. 168). A bijective mapping to a simplicial complex is not always possible. If there are some

---

[9]In practical applications the relation is most often done in a way which puts the points in the *center* of the cells. This is, however, not important here. We only want to show that *there is* a mapping between a point grid and a cell grid. The *interpretation* is, of course, up to the application.

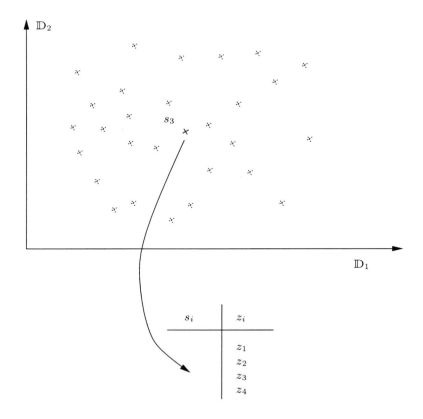

Figure 4.9: Irregular points, dim $\mathbb{D} = 2$

regularities[10] in the set of s_i then there is no related unambiguous Delaunay-triangulation (see section 4.4.3). Figure 4.9 shows an example of irregular points in two dimensions.

4.4.6 Contour models

Contour models can be seen either as a specialization of an irregular-point model or a specialization of a polyhedral tessellation. The contour model is the only one within this group that gives a strong characterization of the associated field measurements $\{z_{i,1}, \ldots, z_{i,m}\}$. In an idealized model, the indices s_i are non-intersecting hypersurfaces in $\mathbb{V}$ (i.e., dim s_i = dim $\mathbb{V} - 1$). The assumption regarding the field $z(\cdot)$ is that $z(\cdot) = z_i \, \forall s \in s_i$ and that $z(\cdot)$ is growing or decreasing monotonic from z_i to z_j on the path p from a point

---

[10]For example, if – in two dimensions – four points are cocircular as in the case of a lattice with orthogonal generator.

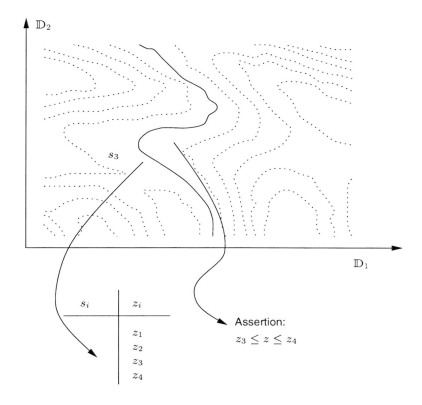

Figure 4.10: Contour line model with dim $\mathbb{D} = 2$

$s \in s_i$ to $t \in s_j$ if p is the shortest path from s to t and does not intersect any other s_k.

However, in practice, the contour models are derived from a set of points defining the contour "line". It can therefore be argued that a contour model consists of a set of irregularly distributed points and that there are subsets in this set of points which are all related to the same field measurement $\{z_{i,1}, \ldots, z_{i,m}\}$. All other interpretations, for example the monotonicity between contour lines, are very dependent on the way the contours were sampled (or digitized) and the measurement context in general. For some applications it is, however, very useful to use additional assumptions on the structure of the field, such as in some interpolation techniques from contour lines (Schneider, 1995). Figure 4.10 shows an example of a two-dimensional contour model.

4.5 Characteristics of the domain $\mathbb{D}$

The discussion of these spatial data models (with the exception of the contour models) has shown that, considering *only* the spatial components s_i of a dataset, there are two classes:

Partition Polyhedral tessellations, with the special cases of cell grids and simplicial complexes, partition the study area $\mathbb{D} \subset \mathbb{A}$ into disjoint indices s_i:

$$\mathbb{D} = \bigcup_i s_i, \quad s_i \cap s_j = \emptyset \quad (4.24)$$

Selection Irregular points and lattices as a specialization select a finite set of points[11] $s_i \in \mathbb{D}$.

However, we have also seen that there is almost always a mapping, e.g., by virtue of a Voronoi diagram or its dual (the Delaunay triangulation), from the *selection* models into a corresponding *partition* model, and vice versa, using, for example., the vertices of the polyhedra or their centers of mass. This means that these models, still considering only the spatial component, are all *equivalent* in the sense that, for example, a cell grid does not carry more information than a lattice. Consider a dataset as given in (4.20) with only one measurement per index s_i:

$$\mathcal{D} = \{M, \{s_i; z_i\}_1^n\}, \quad s_i \subset \mathbb{D}, \quad z_i \in \mathbb{R} \quad (4.25)$$

If there is a one-to-one mapping $A(\cdot)$ from the set of indices $\{s_1, \ldots, s_n\}$ into another set of indices $\{s'_1, \ldots, s'_n\}$ with $s'_i = A(s_i)$, e.g., mapping n nodes of an irregular point model into an n Voronoi polyhedra, then the dataset $\mathcal{D}'$ with

$$\mathcal{D}' = \{M, \{s'_i; z_i\}_1^n\}, \quad s'_i = A(s_i) \quad (4.26)$$

is equivalent to $\mathcal{D}$, i.e., $\mathcal{D}' \equiv \mathcal{D}$.

The objective of this discussion is to show that the choice of the spatial data model is irrelevant *unless* the choice of indices s_i has something to do with the definition of the functional f_i mapping $z(\cdot)$ into z_i. This is, of course, almost always the case. It is therefore sensible to use a spatial data model where the functionals f_i are related to the s_i in a most simple manner, for example, by a selection function $\phi_i(s_i; \cdot)$ which is parameterized by s_i.

The following section will now discuss some of the characteristics of the mapping f_i for the various types of the (spatially equivalent) spatial data models.

---

[11] The notation is not very rigid in the sense that we defined s_i to be a *subset* of $\mathbb{D}$ and not an element of it. Correctly we would have to write $s_i = \{\tilde{s}_i\}$ with $\tilde{s}_i \in \mathbb{D}$.

4.6 Characteristics of the range

The previous sections discussed mainly the spatial component s_i of the various spatial data models. The corresponding values z_i or $z_{i,j}$, respectively, were assumed to be given by a set of functionals f_i and $f_{i,j}$. However, these functionals are almost never known explicitly. They are only a *mathematical model* which defines the relation between spatial index and measured field value. This relation is strongly dependent on the characteristics of all stages in the measurement process. Knowledge regarding the measurement process is therefore necessary to interpret a dataset $\mathcal{D}$, which is basically a collection of numbers. One way of incorporating such knowledge into a dataset is to select a spatial data model for the representation by a spatial structure that optimally reflects the measurement. For example, the measurement of the air temperature on the earth's surface at a selected number of sites is well represented by a model using irregularly distributed points, since it is the interest of the experiment designer to use "localized" instruments, i.e., selection functions ϕ_i with an almost point-like support.

However, there is always a trade-off between computational simplicity and realistic[12] representation. Consider, for example, a satellite image. Here, the measurements deliver an array of numbers produced by the satellite sensor. Each number is associated with a "cell" of the sensor. A sensor which is measuring back-scattered radiation from the earth's surface, for example, measures in each cell an accumulation of the radiation intensity which hits the cell through the sensor's optics. What spatial index can these values be associated with? In a first approximation, one sees that almost all radiation is coming from the earth's surface, and that the radiation intensity (including the atmosphere attenuation) can be related to a region on the earth's surface. If the sensor cell and the optics were perfect, the region on the surface would be the image of the cell on the earth's surface; and, with a square cell, a square projected onto the (curved and rough) earth's surface would be the image along the optical axis. Therefore, a cell grid representation seems appropriate. However, the sensor cell *accumulates* the radiation intensity, i.e., the value produced is an *average* over the cell area, possibly weighted with a maximum in the center of the cell. A user of such a dataset – not aware of the measurement details – might interpret the information that "the dataset represents averaged intensity z_i for each cell s_i" in the sense that for every $s \in s_i$, the average of $z(\cdot)$ in a "neighborhood" is z_i, which is wrong, as can be seen for the values on the boundary of s_i. Would it be better to represent the dataset as a lattice? It certainly would not directly suggest any values "between" the lattice points, but the implicit information over which area the averaging occurred is lost[13].

---

[12] *Realistic* meaning here that the representation carries as much of the information on the relation between values and indices as possible.

[13] Although one would expect that a user of a dataset would assume that the averaging occurred on a region defined by the half-distances to the neighboring lattice points.

The same problems occur in polyhedral tesselations. Often, each polyhedron or polygon s_i is associated with a *categorical* value z_i which was somehow calculated, estimated or measured from the underlying field $z(\cdot)$ with the assumption that z_i is constant over s_i. In many cases, the spatial structure is even determined by categorization (reclassification, contour models), e.g., the polyhedra are constructed using another field representation with a higher sampling density. In these examples, it is also usual for z_i to relate to the entire polyhedron s_i and it cannot be deduced that the category value z_i is the same for every $s \in s_i$[14].

Simplicial complexes provide another example of the danger of inferring information from the spatial data model. In two dimensions, simplicial complexes or triangular irregular networks (TIN) are often used to describe surfaces, e.g., digital terrain models. Each *vertex* of the TIN is associated with a value of the field. It is important to note that this is fundamentally different from the way simplicial complexes were described in section 4.4.3. Here, each *vertex* of a simplex or triangle is a carrier of a value, as opposed to the entire simplex itself. This is basically an *irregular points model*. The triangular structure carries *no explicit* information of the field at all. However, it makes sense sometimes to provide such a triangulated structure together with the individual points because this structure can be used for a subsequent interpolation. Moreover, the TIN might have been generated from a higher-resolution dataset with the goal that the field value $z(s)$ for $s \in s_i$ be best approximated by a predefined, simple function of the values in the vertices of the simplex and the position of s within s_i (e.g., filtering).

The main message here is that it is very important to understand what information a dataset represents. The *implicit information* given by the spatial data model used might support the understanding or be misleading. It is an important guideline for an appropriate selection of a representation technique to *minimize* the implicit information that is needed or which is inherently available in the data model. Positively expressed this means that as much of a dataset's information content as possible should be made explicit.

4.7 Analysis of continuous field data

Given a data set $\mathcal{D}$ as a representation of a field $z(\cdot)$, it is often desired to reconstruct the field from the measured data values, i.e., to estimate a field $\hat{z}(\cdot)$ using the information provided in $\mathcal{D}$. This step is sometimes called *objective*[15] *field analysis* (Hiller & Käse, 1983; Thiébaux & Pedder, 1987; Franke, 1990) or, in the context of spatial information systems, *spatial interpolation*. In this

---

[14]This starts to be even more complicated if the underlying field $z(\cdot)$ is quasi-continuous, i.e., if the notion of continuity is very scale-dependent. A common example is the "field" of population density.

[15]Objective referring here to "objective methods" in contrast to subjective methods that were used, e.g., in meteorology or oceanography to manually draw contour maps from point measurements.

section some of the main characteristics of such techniques are discussed. The objective is not to describe the methods themselves but to give an overview of the kind of operations needed to transform $\mathcal{D}$ into $\hat{z}(\cdot)$.

The primary objective of the analysis of fields is the estimation of a field value $\hat{z}(s_0)$ at a subset s_0 of $\mathbb{D}$. Depending on the distribution of the indices s_i and the set s_0, one might distinguish:

Interpolation: The objective of interpolation is to estimate values "between" sampled values, i.e., to increase the resolution of a currently available information source. $z(s_0)$ is found by interpolation if s_0 is within the convex hull of all indices s_i and if s_0 and all s_i are disjoint, i.e., $s_0 \cap s_i = \emptyset$ for all s_i[16].

Extrapolation: Extrapolation is similar to interpolation except that the location s_0 is *not* within the indices' convex hull.

Aggregation: $z(s_0)$ is determined by aggregation if s_0 and the indices s_i are *not* disjoint, i.e., $s_0 \cap s_i \neq \emptyset$ for some i.

In most cases, the problem statement can be reduced to interpolation. Extrapolation can be seen as a "pathological" special case of interpolation, and aggregation can most often be reduced to a set of interpolation problems. Consider for example the interpolation problem where s_0 is a region of $\mathbb{D}$ and the mean of the field over the region is needed (e.g., block kriging (Isaaks & Srivastava, 1989)). A typical approach is to estimate the field value $z_{0,i} = \hat{z}(s_{0,i})$ at n point locations $s_{0,i} \in s_0$ and then to calculate the mean from these point estimates. It is obvious that, if the distribution of the $s_{0,i}$ in s_0 is uniform, the average will approach the true value for $n \to \infty$. The same approach can also be used for aggregation problems, and thus, the aggregation can be reduced to interpolation[17]. The central problem statement can, therefore, be reduced to the case where s_0 is a point.

---

[16]Interpolation is sometimes required to reproduce the function values z_i "at" s_i in contrast to *approximation* methods. Here we do not, in general, see z_i as the function value of $z(\cdot)$ at s_i, since s_i might be a region and z_i derived by a complicated functional f_i. Approximation usually means in this context to find a function $g(\cdot)$ which "best" possibly approximates a function $f(\cdot)$. Here, we do not know the *true* function $f(\cdot)$.

[17]However, for some applications, it is not *efficient* to reduce the aggregation problem to a set of interpolation problems because, for example, there are sometimes geometric ways to solve it. Consider a population dataset (polyhedral tessellation or polygon coverage, respectively) defining the number z_i of people living within each polygon s_i. To estimate the population on a new polygon s_0 which intersects some of the original polygons s_i, one can use – assuming that the population is evenly distributed over each s_i in the scale-range considered – the sum of the fraction of population given by the part of s_0 that covers each s_i, i.e.,

$$z_0 = \sum_{i=1}^{n} \frac{\|s_0 \cup s_i\|}{\|s_i\|} z_i \qquad (4.27)$$

($\|\cdot\|$ denotes the "volume of" or "area of"). So it is not necessary to sub-sample at many $s_{0,i}$ within s_0.

There are many interpolation techniques available, ranging from very simple linear interpolation to sophisticated geostatistical approaches (Lam, 1983; Isaaks & Srivastava, 1989). The basic objective of these methods is to find a method $\hat{z}(\mathcal{D}; s)$ which minimizes some functional $F(z(\cdot), \hat{z}(\cdot; \cdot))$. This functional might for example express that the absolute difference between the "true" field and the estimated values is minimized within the area of interest, i.e.,

$$F(z(\cdot), \hat{z}(\cdot; \cdot)) = \int_A |\hat{z}(s) - z(s)| ds \stackrel{!}{=} \min \qquad (4.28)$$

The general problem is that the "true" value of $z(\cdot)$ is never known. Therefore, it is necessary to incorporate additional information into the selection or definition of the method, such as:

- Domain-specific knowledge about the phenomenon represented by $\mathcal{D}$. For example, a typical spatial *correlation length* of the phenomenon.

- Stochastic nature of the phenomenon as given by estimating the *autocorrelative structure*.

- Related dataset(s) $\mathcal{D}'$ which are known to be correlated in some sense to $\mathcal{D}$, e.g., for multiple regression techniques or co-kriging (Isaaks & Srivastava, 1989).

The optimization goal is very application-specific. While some applications need a minimum-distance-optimization as given by (4.28), others might require smoothness (minimal curvature or "stress energy").

The selection of an appropriate method is a non-trivial task and depends on many factors (Bucher & Včkovski, 1995). As shown in (Englund, 1990), it is nonetheless very important to use a "good" method for a reliable use of the derived data. The interpolation results are very sensitive to the method selected.

4.8 Impediments

The practical use of datasets describing continuous fields is complicated and offers many problems to data users. These problems are induced by the discretization step involved when mapping $z(\cdot)$ to the discrete set $\mathcal{D}$. This mapping, defined by the functionals f_{ij}, is usually a part of the sampling strategy for the phenomenon under consideration. Ideally, the spatial (and temporal) indices s_i would be selected in a way which makes the data most useful for the application in mind, i.e., for the later use of the data. In most cases, however, there are other factors influencing the choice of functionals f_{ij} or indices s_i, respectively:

- The measurement devices always impose limits on all aspects of the measurement process, e.g., precision, accuracy, spatial and temporal distribution. A satellite sensor, for example, is *by design* limited to data collection on a lattice or cell grid.

- Non-technical factors such as sampling costs and other restrictions limit the measuring intensity, e.g., spatial and temporal density of the measurements.

- Datasets are sometimes collected *without a specific application* in mind. National and international organizations, such as meteorological offices, collect and provide data for various applications. Therefore, the sampling strategy cannot be designed in an application-specific way, but it has possibly to meet the needs of all "interested" applications.

- Reuse of existing datasets is sensible wherever possible. This means, however, that datasets have to be used which were sampled *by a different sampling strategy*. This also applies to cases where the data needed are *derived* from existing data sets rather than by direct measurement.

Owing to these factors, a dataset $\mathcal{D}$ describing a continuous field is rarely used "as is". It needs rather to be transformed to meet specific application requirements. These usually affect both the domain and the range of the field:

Domain Field values are needed for indices s_0 that are not available in the given set of indices $\{s_i\}_1^n$. This means simply that one often needs values at "unsampled locations". The transformations used here are *interpolation methods*[18].

Range The definition of the dataset $\mathcal{D}$ was based on a relation between an index s_i and a set of numbers $\{z_{i,1}, \ldots, z_{i,m}\}$ representing the "field value at s_i". In many cases, a representation $\{z'_{i,1}, \ldots, z'_{i,l}\}$ is needed for a specific application. This might require a simple conversion into another unit system or a change of the uncertainty representation, e.g., a mean value instead of histogram classes.

In general, we can say that a specific application will require the estimation of a new dataset $\mathcal{D}'$ with a set of functionals f'_{ij} different from the functionals defining $\mathcal{D}$.

4.9 Review

For the implementation of VDS and OGIS the following results of this chapter can be summarized. The comparison of various spatial data models has shown that there is no fundamental difference between the data models from a data user's point of view. The *well-defined interface* basically consists of a "function" $z(\cdot)$ which returns a field value for every $s \in \mathbb{D}$. The implementation of $z(\cdot)$ needs to encapsulate the methods that perform the transformations

---

[18]Including extrapolation and aggregation as discussed in section 4.7.

mentioned above. The implementation of these methods, however, will *need* information about the spatial data models in most cases. Availability of information, such as regularities in the spatial ordering, might significantly increase the efficiency of the algorithms. Consider, for example, a spatial search which is needed in most interpolation techniques to determine a set of nearest neighbors. Searching the nearest neighbors on a grid is very simple because of the implicit structure of a grid. Determining the nearest neighbors from a set of irregularly distributed points, on the other hand, needs sophisticated methods to be efficient[19].

The notion of the "function" $z(\cdot)$ together with the domain $\mathbb{D}$ and the range $\mathbb{V}$ can be seen as an *essential model* for a subsequent implementation, that is, a field specified by identification of the triple $(z(\cdot), \mathbb{V}, \mathbb{D})$.

---

[19]Efficiency is necessary if values are to be calculated *on request*.

CHAPTER FIVE

Modeling uncertainties

5.1 Introduction

The previous chapter discussed the mapping of phenomena varying continuously over space and time to mathematical objects, and, eventually a set of tuples in relation (4.19). The presentation was limited to uncertainties caused by *random variations*, i.e., uncertainties that can be modeled using stochastic approaches. In this chapter the modeling of uncertainties is discussed on a more general level, not specifically relating to continuous fields. The objective is to guide the implementation with appropriate uncertainty models, i.e., an appropriate selection of functionals ψ_j. These models will serve as *essential models* for the implementation. For this reason we will step back here and discuss some general characteristics of uncertainty models.

As had been said before (page 67), mathematical models are one of the most important and useful epistemic principles in natural sciences. The tools of mathematics can greatly support scientific insight once a mapping between the real-world phenomenon and corresponding mathematical objects is found. In the past few decades the importance of mathematical models has also grown because these models are a presupposition of translating real-world processes into digital computers. The models can be as simple as assigning a number to a phenomenon, i.e., defining a *quantity*, or defining a deterministic relationship between two or more quantities. Other models use partial differential equations to describe interrelations between several natural processes. It is important to distinguish this broader notion of *model* from computer models used for simulations. Thus, a model is a mathematical description of the corresponding reality, or, mathematics is a language to describe our perception of reality. The mathematical theory dealing with questions of this kind is called *mathematical system theory* (Kalman, 1982; Leaning *et al.*, 1984) and is a rather young discipline within mathematics.

One of the very basic models is the aforementioned assignment of numbers to a natural phenomenon. This assignment is based either on a *measurement*, or, using previously defined models and assignments to numbers, on *prediction*. As an example, let us consider a falling pebble in a constant gravity potential.

One mathematical model to establish is the notion of *velocity v* as a quantity, i.e. a *real number*. The measurement of the velocity yields a number that is interpreted as *velocity*.[1] Another, more complicated model is Newton's law, stating that the rate of change of the velocity is proportional to the force F, which is constant in our example. So we have

$$\frac{d}{dt}v \propto F \tag{5.1}$$

Having a measured velocity v_0 at time t_0 we can predict[2] the value of the velocity at any other time t to be

$$v(t) = \frac{F}{m}(t - t_0) + v_0 \tag{5.2}$$

Unfortunately, the mapping (i.e., the model) of reality to a mathematical object is never one-to-one. The models mentioned above are *idealizations*. All models are affected by uncertainty to some degree. It might be impossible, for example, to consistently assign a number to what we defined to be the velocity. We cannot say that two velocities are the same if the numbers given by the measurements are the same, nor will two measurements of a constant velocity yield the same numbers. And models defining relationships (e.g., (5.1)) are constrained by our limited scope of perception and are therefore uncertain to some degree.[3] As a consequence, all inferences drawn from such measurements and models will be affected by uncertainties. But what *are*

---

[1] The real process of measurement implies a lot of other models of course, but for the sake of simplicity let us assume that there is a way of just looking at the falling pebble and getting an idea of a number to associate with the notion of velocity. System theory calls the measurement process yielding a number the *behavior* of the system.

[2] Of course, we also need models of *time*, *acceleration* and so on.

[3] Such models cannot be *proven* rigidly. They are always based on some assumptions. One of those assumptions is the epistemic principle to choose the *simplest* and *most reasonable* model of a family of valid models, i.e., of a family of models compatible with our observations. *Simple* and *reasonable* in the field of natural sciences might be interpreted as "mathematically easily tractable", "a model that we understand", "elegant" and so on, i.e., the choice is *subjective*. In mathematics it is possible to rigidly prove some statements based on an unprovable axiomatic frame. The *simple* and *reasonable* "choice of models" corresponds to the selection of a set of axioms. A set of axioms is *Simple* and *reasonable* if a lot of true statements can be derived from it, and, if no axiom can be derived from the other axioms in the set (this is actually the definition of *axiom*). But even then, it is impossible to prove each true statement that is compatible with the axiomatic system. This is an essential inference of Gödel's famous theorem (Nagel & Newman, 1992).

those uncertainties? How can we cope with the – quite obvious – statement that "everything is uncertain to some degree"? Is there a way of characterizing or even measuring a "degree of uncertainty"? These questions are not at all new to the scientific community. Scientists have been engaged with similar problems for a long time and there are well-known disciplines concerned with uncertainties such as statistics and probability theory (Barlow, 1989), measurement theory (Destouches, 1975; Narens, 1985; Kyburg, 1992) and all applied sciences as well.

This topic gets a very practical touch in the context of digital processing of scientific data. The soft- and hardware environments used are often not capable of dealing with uncertainties in the same manner as domain specialists do. Consider for example the digital representation of numbers with a limited precision. There is a convention in the scientific community telling us to communicate *only the significant digits* of a measured quantity. If we say "the temperature value is $T_1 = 5.1$ centigrade" we implicitly say that the precision of the measurement is better than 0.1 centigrade and that we believe that the true value T lies in the range $5.05 \leq T < 5.15$. If the measurement had been precise to 10^{-3} centigrade we would write $T_2 = 5.100$. This convention is simple and useful, although it is based on a decimal system and only allows precision to be represented on a logarithmic scale (10^{-n}). Digital computers do not follow this convention. A digital computer representing the values $T_1 = 5.1$ and $T_2 = 5.100$ using one of the standard formats (i.e., IEEE floating point representation (IEEE, 1987)) *cannot distinguish* T_1 and T_2. $T_1 - T_2$ yields zero even though the true value might be somewhere in the interval $[-0.05, 0.05]$. Sensible *mathematical* models for uncertain values are therefore also useful for the *digital* representation of the values.

This chapter will review some mathematical methods for handling uncertainties. The next section tries to provide an understanding of the term "uncertainty". Some kinds of uncertainty are discussed together with some common concepts of the methods presented in section 5.3. Several basic methods for mathematically modeling and exploring the impact of uncertainties are compared, such as:

- *Probability theory:* the classical way of modeling uncertainties.

- *Interval methods:* for easy and fast computations.

- *Fuzzy methods:* a generalization of interval methods suitable for the representation of vague information.

- *Dempster–Shafer theory of evidence:* a theory for quantitative reasoning under uncertain conditions covering both probability theory *and* fuzzy methods as special cases.

- *Perturbation methods, Monte Carlo techniques* and *Gaussian error propagation:* assess the effects of uncertainties on functional relationships and their *sensitivity*.

The methods are compared with respect to their usability and range of application. The last section will discuss some issues regarding the implementation of these models on digital computers.

5.2 Uncertainties

5.2.1 What is uncertainty?

In the introduction we introduced the notion of "uncertainty" without defining its meaning. Bandemer and Näther (1992) distinguish three major types of uncertainty:

Variability The system under consideration can be in various states under similar conditions. This is an issue of *chance* or missing information. As in the case of imprecision below, it is not important for our discussion whether the variability is due to an incomplete understanding of the system (e.g., "hidden variables" (Mermin, 1985)) or whether the system is inherently non-deterministic to some degree.

Imprecision It is impossible to observe or measure to an arbitrary level of precision. In classical measurement theory one sometimes distinguishes between *accuracy* and *precision* (Piotrowski, 1992). The first describes the error of a measurement, the latter its "crispness", or – using the terminology of mathematical statistics – bias and variance. It is not important for our discussion whether the limits of precision are inherent and therefore never can be overcome, or if these limits are of a more practical nature.[4] Even if we believe in theoretically unlimited precision, practical issues always will limit the reachable precision.

Vagueness This is mainly due to descriptions using natural language. As an example, classifications during data collection using rules written in a natural language depend on the observer's interpretation of the rule. Another very typical example of vagueness is propositions stated by experts. Experts will often express their opinions with a relatively high degree of vagueness, since they have their own experienced understanding of what "high", "strong" and "very low" mean in the context under consideration.

These types of uncertainty can and usually do occur in combination. Uncertainties can be viewed as a sort of information deficiency (Klir, 1994), even if their presence is fundamental and not due to missing information. This simplification can be justified as follows. We are interested in the analysis and modeling of uncertainty *based on observations*. It is very difficult, if not impossible, to decide from observations alone whether some uncertainties (e.g.,

---

[4]This is a fundamental question with the interpretation of quantum mechanics (Busch *et al.*, 1991).

variabilities) are a matter of chance (non-deterministic) or if they *seem* to be random (but are deterministic) because we do not know enough about the process observed.[5]

5.2.2 Basic concepts of modeling uncertainties

In the introduction we described the concept of *mathematical model* as an epistemic principle. This principle is based on the assignment of real-world properties to mathematical objects. We can define the real-world properties under consideration as *propositions* and their mathematical counterpart to be *sets*. A simple value assignment (e.g., measurement) is the proposition A "the surface air temperature is $20°$ centigrade", or, equivalently, $T_0 = 293K$. A functional relationship is the proposition B "the vertical temperature gradient (lapse rate) is nearly constant to $0.5°$ centigrade per 100 meters", or, $\gamma \approx 5 \cdot 10^{-3} \frac{K}{m}$. Together with the proposition C "the temperature at height h above the surface is found by multiplying the height with the lapse rate γ and adding the surface temperature T_0", or $T(h) = T_0 - \gamma h$, we can imply D "the temperature 200 meters above surface is $19°$ centigrade":

$$A \cap B \cap C \subset D \tag{5.3}$$

This example is not very rigid, as for example proposition C is in fact the same as A and B, since C is the direct implication from the definition of "(surface) temperature", "gradient" and "constant". This demonstrates, however, a way of modeling statements about reality via mathematical objects.[6] The handling of these objects, or sets, requires some basic definitions[7]:

- proposition: $\theta \subset \Theta$
- intersection or conjunction: $\theta = \theta_1 \cap \theta_2$
- union or disjunction: $\theta = \theta_1 \cup \theta_2$
- complement or negation: $\theta = A^c$

---

[5]The interpretation of quantum theory concerning the fundamental question of determinism continues to be discussed. It is, incidentally, very interesting to see that those discussions about objectification etc. have grown in the past decades. In the middle of this century the scientific community started to *believe* in the new basic theories of nature, and criticism such as the famous paper of Einstein, Podolsky and Rosen was refused, e.g., (Busch *et al.*, 1991). During the past decades, however, it has been realized that some issues of quantum mechanics are not at all clear and that there are several possible interpretations.

[6]It is important to note that the mathematical objects are the sets A, B, C and D, and *not* the formulas like $T(h) = T_0 + \gamma h$. It is nonetheless usual to view such a formula as a mathematical object, see footnote 7.

[7]The notation here is somewhat sloppy. The representation of propositions as *sets* can be derived thoroughly, e.g., (Shafer, 1976). In the following we will silently assume that the reader is aware of this isomorphism. The benefit of this view lies in the simple representation of (measurement) values which are known to be from a set Ω and the representation of relationships between those values. Then we can define Ω' as the set of all propositions "the value is ω" for all $\omega \in \Omega$. Ω' is therefore isomorphic to Ω.

- empty set or false proposition: $\theta = \emptyset$
- whole set or (always) true proposition: $\theta = \Theta$
  Θ is called *frame of discernment* by some authors (Shafer, 1976).[8]

Using this framework we can now give two definitions of uncertainty. These definitions are strongly related but not equal. They are based on the use of "uncertainty" in the natural language. As an example let us consider two uncertain propositions about a possible rise in the mean annual temperature $\Delta\overline{T}$ after 50 years.

$\theta_1 =$ "$\Delta\overline{T}$ might be $1°$"

Definition 1 *A proposition θ is uncertain if it is neither true nor false.*

$\theta_2 =$ "$\Delta\overline{T}$ is in the range $[-10°, +10°]$ for sure"

Definition 2 *A true proposition $\theta_2 \subset \Theta$ is uncertain if and only if there are subsets ω_1, ω_2 of θ_2 ($\omega_i \neq \emptyset$, $\omega_i \subset \theta_2$, $i = 1, 2$) with $\omega_1 \neq \omega_2$. If Θ is finite, this means that $|\theta_2| > 1$ (θ is not a "singleton").*

Using these definitions we can make the following observations:

- A proposition is *not uncertain* only if it is true and does not contain any proper subsets. The notion of *certainty* is not usually the same, i.e., it is not an inversion. Definition 2 is not appropriate for certainty since one often interprets certainty with "truth". θ_2 is certain because it is true, and uncertain because it is ambiguous. This is one of the main reasons for a lot of confusion when talking about uncertainty. When used in the sense of definition 2, it might be more appropriate to use "imprecision" or "ambiguity".

- An uncertain proposition in the sense of θ_1 implies a true proposition θ_2 (which is uncertain or ambiguous respectively), i.e., there is always a true θ_2 with $\theta_2 \supset \theta_1$. This means that it is possible to reduce uncertainty at the expense of unambiguity.

- The set of all propositions of Θ is its power-set[9] 2^Θ.

- Both definitions show that we need some kind of measure and notion of the *belief*, "possibility" or "probability" that a proposition is true. This might be either two-valued (true, false), or other finite or infinite sets of values expressing the belief in or the probability of a proposition. If

---

[8] Θ might be finite or infinite, but for the sake of simplicity, we assume (at least for this section) that Θ is finite.

[9] The power-set 2^M of a set M is the set of all its subsets, i.e. $2^M = \{m | m \subset M\}$. The notation of 2^Θ for the power-set is very intuitive for finite sets, since the cardinality of the power-set is given by: $|2^\Theta| = 2^{|\Theta|}$.

we use a subset of the real numbers to express the "degree of truth" it can be defined as a function $\mathcal{D}_T(\theta) : 2^\Theta \to \Pi$ with $\Pi \subset \mathbb{R}$.[10]

- To make use of definition 2 a measure of *ambiguity* would be useful.

5.2.3 Inference from uncertain propositions

The previous section showed a formalization of propositions and a need to have a corresponding measure for the "degree of truth". In order to allow inferences from given propositions the set theoretic interpretation of propositions can be enhanced using basic set operations of conjunction, disjunction and negation. A useful model of uncertainty therefore should allow the calculation of $\mathcal{D}_T(\theta_1 \cup \theta_2)$ and $\mathcal{D}_T(\theta_1 \cap \theta_2)$ given $\mathcal{D}_T(\theta_1)$ and $\mathcal{D}_T(\theta_2)$ and possibly some other parameters, i.e.:

$$\mathcal{D}_T(\theta_1 \cup \theta_2) = f_\cup(\mathcal{D}_T(\theta_1), \mathcal{D}_T(\theta_2), \ldots) \qquad (5.4)$$
$$\mathcal{D}_T(\theta_1 \cap \theta_2) = f_\cap(\mathcal{D}_T(\theta_1), \mathcal{D}_T(\theta_2), \ldots) \qquad (5.5)$$

This allows us to calculate the "degree of truth" of a proposition derived from propositions with a known "degree of truth". A very useful example is the "error propagation problem".

Let A and B be uncertain (and ambiguous) measurements, both of them being propositions of $\Theta = \mathbb{R}$. A and B shall be decomposed into a set of disjunct possible "values" a_i and b_i, i.e. $A = \bigcup_i a_i$, $a_i \cap a_j = \emptyset$ for $i \neq j$. Furthermore, we shall know the "degrees of truth" $\mathcal{D}_T(a_i)$ and $\mathcal{D}_T(b_i)$ for every a_i, b_i. If we want to assess a derived value C which is given by $C = f(A,B)$ we can use basic set algebra:

$$C = \bigcup_i \bigcup_j c_{ij} \qquad (5.6)$$

where

$$c_{ij} := f(a_i, b_j) \qquad (5.7)$$
$$= \{c \in \Theta \mid c = f(a,b) \land a \in a_i \land b \in b_i\} \qquad (5.8)$$

Using the rules from (5.4), $\mathcal{D}_T(c_{ij})$ and thus $\mathcal{D}_T(C)$ can be calculated. As an example consider $A = [0,2] = a_1$, $B = [3,4] = b_1$ and $f(A,B) = A + B$, $\mathcal{D}_T(a_1) = d_a$, $\mathcal{D}_T(b_1) = d_b$. Then $c_{1,1}$ is the set of all numbers that are the sum of the numbers in $[0,2]$ and $[3,4]$. Since f is monotonic it is easy to see that $c_{1,1} = [3,6]$. With $C = c_{1,1}$ we have $\mathcal{D}_T(C) = f_\cup(d_a, d_b, \ldots)$. Some approaches for defining measures $\mathcal{D}_T(\cdot)$ and their relations are presented in the next section.

---

[10]In the next section some approaches for defining such functions are presented. Those functions have names like "probability", "possibility", "necessity", "belief" and so on. These words have specific meanings in the corresponding theories and it is therefore dangerous to use them without those definitions.

5.3 Methods for the modeling of uncertainty

5.3.1 Overview

The following presentation of some of the most widely used modeling techniques for uncertain information is intended to reveal the basic concepts and motivations behind the various approaches to uncertainty and their similarities and differences. The previous section introduced uncertainty as a statement about *propositions* in a very general way. It is obvious that appropriate methods strongly depend on the kind of uncertainty faced. Dealing with measurement errors is quite different from the modeling of vague expert knowledge (Hofmann, 1975; Finkelstein & Carson, 1975). Incorporating partial ignorance might demand methods other than controlling rounding errors on digital computers. And the influence exercised by uncertainties on a system of non-linear differential equations is analyzed by methods other than vague statements in a questionnaire. Nonetheless, there *are* similarities between the various approaches, and they can be shown using the formal approach introduced in the previous section. The presentations are by no means complete and are only intended to give a short overview and point to other, more specific literature.

5.3.2 Probability theory

Probability theory, or stochastics, is certainly the most widely used way of modeling uncertainty. It has its origins in the sixteenth century when humans tried to better understand games of chance. The concept of chance is likewise very important in stochastics. A rigid axiomatic frame for probability theory was introduced in 1933 by A.N. Kolmogorov (Kolmogorov, 1933). It is based on tree properties of a "degree of truth"-measure $\mathcal{P}_{rob}(\cdot) : 2^{\Theta} \to [0,1]$ called *probability*:

$$\mathcal{P}_{rob}(\theta) \geq 0 \qquad (5.9)$$
$$\mathcal{P}_{rob}(\Theta) = 1 \qquad (5.10)$$
$$\mathcal{P}_{rob}(\theta_1 \cup \theta_2 \cup \ldots \cup \theta_n) = \mathcal{P}_{rob}(\theta_1) + \mathcal{P}_{rob}(\theta_2) + \ldots + \mathcal{P}_{rob}(\theta_n)$$
$$\text{if } \theta_i \cap \theta_j = \emptyset \, \forall i \neq j \qquad (5.11)$$

In stochastics, the propositions θ are called *events*. From these axioms[11] many simple statements can be derived. Important for our discussion are:

$$\mathcal{P}_{rob}(\emptyset) = 0 \qquad (5.12)$$
$$\mathcal{P}_{rob}(\theta^c) = 1 - \mathcal{P}_{rob}(\theta) \qquad (5.13)$$
$$\mathcal{P}_{rob}(\theta) \leq 1 \qquad (5.14)$$

---

[11] It should be noted that for an infinite Θ the axioms have to be slightly modified. Instead of the 2^{Θ} one defines the mapping to be $\mathcal{P}_{rob}() : B(\Theta) \to \mathbb{R}^+$, where $B(\Theta)$ is a θ-induced Borel-field and the axiom of addition (5.11) is enhanced to support "infinite sums".

The first one states that the probability of the proposition known to be false is zero. The second (5.13) is a direct consequence of the addition axiom (5.11) and states that the probability of the negation of a proposition θ is 1 minus the probability of θ. In other words, the proposition "it is either θ or not θ" is absolutely true. This is intuitive in many situations if we know what Θ, the "frame of discernment" is. If Θ is the simple set {"the ball is red", "the ball is green"}, then it is perfectly reasonable to trust the proposition "the ball is either red or not red", because we known that if the ball is not red then it is green and there are only red and green balls. But there are often situations of (partial) ignorance where Θ is not exactly known. It might be wise not to fully trust the statement "the ball is either red or not red" if we suspect that there might be other balls as well. Such situations of ignorance are dealt with in the Dempster–Shafer theory of evidence presented later on in section 5.3.5.

Probability theory is only useful if there is a consistent way to give values to the probability function $\mathcal{P}_{rob}(\cdot)$. The importance of stochastics stems from the so-called *frequency interpretation* which gives sich a way of estimating $\mathcal{P}_{rob}(\cdot)$ based on real-world observations. This interpretation is also the foundation of mathematical statistics, being one of the most widely used ways of doing data analysis. Frequency interpretation[12] means that the relative frequency of m occurrences of an event θ during n observations approaches the probability of θ if n is large:

$$\mathcal{P}_{rob}(\theta) \approx \frac{m}{n} \qquad (5.15)$$

It is essential in this interpretation to assume that observations are *repeatable* under identical circumstances.[13] The issue of *repeatability* is another frequently mentioned criticism of the application of stochastics for certain problems.

If Θ is $\mathbb{R}$, as it is usually when measuring physical quantities, one can define a *probability distribution function* with

$$F(x) = \mathcal{P}_{rob}(X < x) \qquad (5.16)$$

for a random variable X. $F(\cdot)$ is usually easier to handle than $\mathcal{P}_{rob}(\cdot)$. There are a lot of well known measures for the properties of $\mathcal{P}_{rob}()$ or $F(\cdot)$ like *expected value, variance, median* and so on. Mathematical statistics gives methods to estimate those measures using observed data. Probability theory and statistics is therefore a tool of major importance in the analysis of measured data in natural sciences.

Useful measures for ambiguity are either the abovementioned variance (and the standard deviation as its statistical counterpart) or for example the *Shannon-entropy*:

$$H = -\sum_{\theta \in \Theta} \mathcal{P}_{rob}(\theta) \log \mathcal{P}_{rob}(\theta) \qquad (5.17)$$

---

[12] This is Bernoulli's version of the central limit theorem.
[13] Technically spoken the observations are realizations of *independent identically distributed* random variables.

The entropy H was introduced in information theory (Shannon, 1948) and leads us to another view of the problem of incorporating ignorance into probability. Let us consider as an example a frame of discernment Θ consisting of just two elements $\{\theta_1, \theta_2\}$. If we have no idea at all about this system, the *maximum entropy principle* yields the same probability for both events, i.e.

$$\mathcal{P}_{rob}(\theta_1) = \mathcal{P}_{rob}(\theta_2) = \frac{1}{2} \tag{5.18}$$

If we describe the same situation with three possible propositions $\{\omega_1, \omega_2, \omega_3\}$, where θ_1 corresponds to $\{\omega_1, \omega_2\}$ and θ_2 to ω_3, then we would say

$$\mathcal{P}_{rob}(\omega_1) = \mathcal{P}_{rob}(\omega_2) = \mathcal{P}_{rob}(\omega_3) = \frac{1}{3} \tag{5.19}$$

and

$$\mathcal{P}_{rob}(\omega_1 \cup \omega_2) = \frac{2}{3} \neq \mathcal{P}_{rob}(\theta_1) \tag{5.20}$$

This not very intuitive consequence is clearly due to the probability theory's requirement that Θ has to be known. In the field of Bayesian reasoning (Pilz, 1993) probability distributions often have to be estimated under situations of partial ignorance. This is one of the criticisms of Bayes theory in its application to decision support.

5.3.3 Interval methods

Interval mathematics is a simple yet powerful way of dealing with real-valued data. The basic idea is to represent entities as closed convex subsets of the set of real numbers, e.g., as intervals within $\mathbb{R}$ or $\mathbb{R}^n$. This interval represents all possible values of that entity. There is no measure associated with the intervals, so we can only deal with uncertainties of type 2, i.e., ambiguities. While intervals were used for a long time to specify a set of possible values (e.g., confidence intervals in mathematical statistics), its mathematics only started to be developed in this century, e.g., (Moore, 1966; Bauch *et al.*, 1987; Mayer, 1989; Moore, 1992; Polyak *et al.*, 1992). An entity A is represented by a 2-parametric set:

$$\begin{align}
A &= [\underline{a}, \overline{a}[\tag{5.21} \\
&= \{a \mid \underline{a} \leq a < \overline{a}\} \tag{5.22} \\
\underline{a} &= \inf(A) \tag{5.23} \\
\overline{a} &= \sup(A) \tag{5.24}
\end{align}$$

Interval arithmetic operators $\diamond \in \{+, -, \times, \div\}$ between intervals (and functions on intervals) can be defined with the following simple correspondence principle. The resulting interval C of an operation $A \diamond B$ contains all results of the arithmetic operation $\diamond$ between all elements $a \in A$ and $b \in B$, i.e.:

$$C = A \diamond B := \{c \mid c = a \diamond b \wedge a \in A \wedge b \in B\} \tag{5.25}$$

or more generally:

$$C = f(A_1, A_2, \ldots, A_n)$$
$$:= \{c \mid c = f(a_1, a_2, \cdots, a_n) \wedge a_1 \in A_1 \wedge \ldots \wedge a_n \in A_n\} \quad (5.26)$$

With this principle one can derive simple rules for arithmetic operators and standard functions, e.g., $\sin(\cdot)$, $\log(\cdot)$. The idea then is to replace all quantities within any calculations with intervals yielding intervals as results and therefore some estimate of the resulting uncertainty. If the rules are properly defined, then it can be guaranteed that the result of an operation lies in the resulting interval. Unfortunately, it turns out that there are some limitations to that approach:

- The validity of some basic properties is limited:
    - There is no distributivity law ($a(b+c) = ab + ac$). It is replaced by the so-called semi-distributivity: $A(B+C) \subseteq AB + AC$.
    - Proper intervals (i.e., $\underline{a} < \overline{a}$) do not have multiplicative and additive inverses, i.e. there is no interval X with $A + X = 0$ and $A \cdot X = 1$.
    - If $0 \in A$ then usually $A - A \neq [0, 0]$, and equivalently if $1 \in A$ then $A/A \neq [1, 1]$.

- It is not always possible to define rules using the correspondence principle. The set C defined by (5.26) is not always an interval. Consider for example the interval extension of the function f:

$$f(x) = \begin{cases} -1 & x < 0 \\ 1 & x \geq 0 \end{cases} \quad (5.27)$$

If we apply the definition of C for any proper interval A with $0 \in A$ then C will be the two point set $\{-1, 1\}$ and not an interval. Therefore, the correspondence principle is often expanded to define the interval extension as the *convex hull* of C, i.e., the smallest interval containing C.

- Owing to the above reasons, intervals may grow rapidly in some cases, i.e. the resulting intervals contain *more* than all "possible values", but it is guaranteed that they *contain* the possible values. It is therefore a conservative way of modeling uncertainties in that intervals are usually overestimated by the arithmetic rules.

- The convex hull of the results of some operations is $\mathbb{R}$ itself, e.g. if $0 \in B$ then $C = A/B = (-\infty, +\infty)$. It is, however, possible to enhance the intervals with:

$$A = \begin{cases} [\underline{a}, \overline{a}] & \underline{a} \leq \overline{a} \\ (-\infty, \overline{a}] \cup [\underline{a}, +\infty) & \underline{a} > \overline{a} \end{cases} \qquad (5.28)$$

- The intervals do not give any informations about the "degrees of membership", e.g., every value of the interval is possible. The maximum-entropy principle, i.e., the absence of any information on the "distribution" of the values within the intervals, would lead to a uniform probability distribution over the interval, i.e. a constant probability density of $1/(\overline{a} - \underline{a})$ within $[\underline{a}, \overline{a}]$.

These limitations and in particular the overestimation of intervals have to be considered when uncertainty is to be modeled using intervals. There are however a lot of very useful features of interval mathematics, too:

- The so-called "monotonicity of inclusion" *guarantees* that the possible results lie within the resulting intervals. This has a particularly useful application in the field of scientific computing: interval arithmetic can be used to control roundoff errors when performing floating point calculations on digital computers. A real number is represented by the machine interval containing this number. A machine interval is an interval with lower and upper bounds that are *exactly* representable within the computer. Substantial rounding errors are very frequent but rarely noticed. Especially when problems are *ill-posed* (e.g., non-linear) or *ill-conditioned* (e.g., several orders of magnitude present) one might experience severe rounding errors. Humans usually trust computers where they should not (e.g., in floating point operations) and they do not trust computers elsewhere (e.g., saving documents on disk). It is surprising, however, that there are few software packages and floating point processors offering enclosure methods (interval methods) to solve standard numerical problems.

- Interval methods are efficient on digital computers. They allow a simple and useful tracking of "ambiguities", i.e., uncertainties of type 2. The computational complexity grows typically by a factor of 2–4 if interval operations are performed. They are also efficient from a storage point of view: an interval is represented by two machine numbers, so the storage requirements grow roughly by a factor of 2.

- Much research is done in this field and there are meanwhile many useful extensions, such as methods to estimate the set of possible solutions when integrating partial differential equations with interval-valued initial or boundary conditions (Baier & Lempio, 1992) and general extensions to the theory (Dimitrova *et al.*, 1992; Polyak *et al.*, 1992; Neumaier, 1993).

- In some applications in natural science the "range of possible values" seems to be more important information than distribution details within that range. When studying ecosystems one might be concerned with the range (i.e., minimum and maximum) of possible air temperatures rather than mean temperatures.

- There are specialized programming languages and class libraries available which support interval mathematics, e.g., for Pascal (Geörg et al., 1993) and C++ (Wiethoff, 1992).

There are approaches which use interval methods with a probabilistic interpretation. Given a probability distribution over $\mathbb{R}$, an interval – being a subset of $\mathbb{R}$ – has a certain probability associated with it. A set $A_1, \ldots, A_n$ of disjunctive intervals ($A_i \cup A_j = \emptyset$ $\forall i \neq j$) and associated probability values $\mathcal{P}_{rob}(A_i) = p_i$ with $\sum p_i = 1$ is called *histogram*. With an appropriate number n of intervals the probability distribution $\mathcal{P}_{rob}(\cdot)$ can be approximated. The interval arithmetic rules can be extended to histograms (*histogram arithmetic*; see (Gerasimov et al., 1991)), but this has to be done carefully, since for many functions $f(a, b)$ extended to intervals, the naive approach $C = f(A, B)$ and $\mathcal{P}_{rob}(C) = \mathcal{P}_{rob}(A)\mathcal{P}_{rob}(B)$ fails.

5.3.4 Fuzzy sets

Fuzzy set theory was founded 1965 by L.A. Zadeh (Zadeh, 1965). Ideas to break up the two-valued classic logic and allow states other than "true" and "false" were discussed in general philosophical contexts and also in mathematics long before, but Zadeh built up a consistent mathematical theory which meanwhile gained a lot of attention, especially in engineering. The motivation of fuzzy set theory is the fundamental proposition of set theory stating that an element either belongs to a set or it does not. There are many situations where one would like to express the situation of an element *partially* belonging to a set, or, to express partial ignorance about the membership of an element in a set. Fuzzy set theory captures these situations by defining a *degree of membership* to a set, which is usually a real number between 0 and 1. If $A \subset \mathbb{M}$ is a "crisp", i.e. normal, set, then A is equivalent to a membership function m_A defined by:

$$m_A : \mathbb{A} \to \{0, 1\} \quad m_A(x) = \begin{cases} 1 & x \in A \\ 0 & x \notin A \end{cases} \tag{5.29}$$

A *fuzzy set* is defined with a function

$$m_A : \mathbb{A} \to [0, 1] \quad m_A(x) \in [0, 1] = \text{degree of membership of } x \tag{5.30}$$

Important relations are defined[14] in analogy to classical set theory:

---

[14] There are actually many different and consistent ways of defining these relations, see (Bandemer & Gottwald, 1993, p. 39).

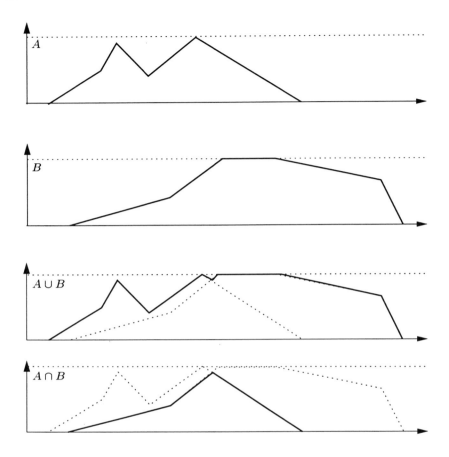

Figure 5.1: Union and intersection of two fuzzy sets

- intersection $A \cap B$: $m_{A \cap B}(x) = \min \{m_A(x), m_B(x)\}$ $\forall x \in \mathbb{A}$
- union $A \cup B$: $m_{A \cup B}(x) = \max \{m_A(x), m_B(x)\}$ $\forall x \in \mathbb{A}$
- complement A^c: $m_{A^c}(x) = 1 - m_A(x)$
- subset $B \subset A$: $m_B(x) \leq m_A(x)$ $\forall x \in \mathbb{A}$

Figure 5.1 shows the union and intersection of two fuzzy numbers, i.e., fuzzy sets on $\mathbb{A} = \mathbb{R}$, as an example. From these definitions it is possible to derive rules for the "arithmetics" of fuzzy numbers. These rules are similar to the previously discussed interval arithmetics, but differ in the important fact that intervals[15] are normal ("crisp") sets.

---

[15]One defines the so called α-cut of a fuzzy set as the crisp set defined with $m_A(x) > \alpha$. For a family of fuzzy sets their α-cuts are intervals, i.e. convex. The arithmetics of such

There is nothing comparable to the frequency interpretation of probability theory in fuzzy set theory. This means that it is not trivial to define the membership function $m_A(x)$ for a given base set $\mathbb{A}$. It is clear (or an intuitive convention) that one defines $m_A(x)$ to be 1 if we are sure that x is in A and 0 if we are sure that it is not. But how to define $m_A(x)$ for cases where we do not know? The rules for this transition zone read often as "to find a full form of m_A, we have to bridge the 'grey-zone' by a suitable monotonic function, which should be as simple as possible and as expressible as necessary" (Bandemer & Näther, 1992). That is – lacking a "natural principle" – the definition of m_A is strongly subjective. This is on the one hand the main conceptual problem when applying fuzzy numbers to represent uncertainties in natural sciences. On the other hand it is very convenient to express expert's statements and other types of vagueness. An *objective* principle inherently inhibits a suitable modeling of such vagueness.

The membership function $m_A(x)$ can be used to define a measure of the "degree of truth" $\mathcal{D}_T$ (). In fuzzy set theory one defines (Bandemer & Gottwald, 1993, p. 147) a *possibility measure* and a *necessity measure* with (if $\mathbb{A}$ is finite):

$$\mathcal{P}_{oss}(\{x\}) = m_A(x) \tag{5.31}$$
$$\mathcal{P}_{oss}(A) = \max_{x \in M} m_A(x) \tag{5.32}$$

and

$$\mathcal{N}_{ec}(A) = 1 - \mathcal{P}_{oss}(A^c) \tag{5.33}$$

Equation (5.33) is the intuitive duality that a proposition is necessary if its negation is not possible. So necessity is a kind of lower limit and possibility an upper limit. It seems intuitive that "probability" is somewhere between those measures. In fact, the Dempster–Shafer theory of evidence discussed below introduces a more general concept yielding probability measures $\mathcal{P}_{rob}(\cdot)$ on the one hand and – under different conditions – possibility and necessity measures on the other.

5.3.5 Dempster–Shafer theory of evidence

The theory of evidence was created in 1976 by G. Shafer (1976) based on earlier work of A.P. Dempster on Bayesian reasoning using "upper and lower probabilities". Shafer extended these ideas and built a consistent mathematical theory.

The theory of evidence is based on so-called *belief functions* $\mathcal{B}_{el}$ () over Θ, $\mathcal{B}_{el}() : 2^{\Theta} \to [0, 1]$. They satisfy:

$$\mathcal{B}_{el}(\emptyset) = 0 \tag{5.34}$$
$$\mathcal{B}_{el}(\Theta) = 1 \tag{5.35}$$

---

fuzzy numbers turn out to be equivalent to interval arithmetics.

$$\mathcal{B}_{el}(\theta_1 \cup \ldots \cup \theta_n) \geq \sum \mathcal{B}_{el}(\theta_i) - \sum_{i<j} \mathcal{B}_{el}(\theta_i \cap \theta_j) + - \ldots$$
$$\ldots + (-1)^{n+1} \mathcal{B}_{el}(\theta_1 \cap \theta_2 \cap \ldots \cap \theta_n) \quad (5.36)$$

As is the case with the possibility and necessity measures, there is a dual measure called the *plausibility* defined by

$$\mathcal{P}_l(\theta) = 1 - \mathcal{B}_{el}(\theta^c) \quad (5.37)$$

The belief functions can be derived from a so-called *basic probability assignment*. This defines probabilities (or *probability masses* as Shafer calls them) for some subsets $\theta \in 2^\Theta$. The difference from classical probability theory, expressed in (5.36), is that the probability is not known for *every subset of* Θ. In the "stochastic limit", that is when we know the *basic probability assignment* for every element of Θ, the belief measure and the plausibility measure coincide with the probability. If on the other hand the sets for which there is a probability assignment are ordered by set inclusion, then one obtains the probability and necessity measures as special cases of the belief and plausibility (Klir, 1994).

The theory of evidence has found many applications in the field of artificial intelligence and decision support systems. It is very general in the sense that it is an extension of classical probability theory. But like fuzzy set theory it suffers from the specification problem. How can we get a belief function for a real problem? The lack of an objective procedure for using observations as a foundation for belief functions makes it not very useful when dealing with "real data" within natural sciences. Nonetheless, every measurement, every mathematical model has to be *interpreted* sometimes. Maybe it has also to be used for decision support and policy making. Especially in the field of research on climate change one is often faced with questions about the impact of scientific results on policy making. How can the scientists inform the public of their results, giving an impression of the uncertainty of their statements? It is worthwhile studying the theory of evidence in order to search for ways to bridge information from pure "exact" natural science into human decision making.

5.4 Assessing the uncertainty of functional expressions

5.4.1 Overview

Within a digital computing environment it is often desired not only to represent uncertain values but also to estimate the resulting uncertainty of functional expressions involving uncertain values. Given a functional expression

$$y = f(x_1, x_2, \ldots, x_n), \quad (5.38)$$

what is the uncertainty associated with the result y? In the context of spatial information processing such questions are very important since typically there are many functional relationships of the form (5.38) involved in most processing steps. Moreover, the user interaction is – by design – often kept on an abstract level, e.g., invoking a specific processing option without knowing what exactly is going on within the software system. It is therefore necessary that the software system itself keeps track of the propagation of uncertainties throughout the transformation to derived data. Unfortunately, the systems currently in use do not automatically honor uncertainty information in functional expressions, mostly because uncertainty information is not even used in the data representation. A notable exception is the *ADAM* software package (Heuvelink, 1993) which was also embedded into a raster-oriented GIS called *PCRaster* (Wesseling et al., 1996).

We will here briefly present some of the approaches for the evaluation of expressions (5.38), which are feasible for an implementation on digital computers.

5.4.2 Gaussian error propagation

Gaussian error propagation is based on the following assumptions (Kamke & Krämer, 1977; Schulin et al., 1992):

1. The uncertainty of each x_i has a stochastic characteristic and is the result of many small independent errors. Each of the x_i can then be assumed to be normally distributed (due to the central limit theorem) with mean μ_i and variance σ_i: $X_i \sim \mathcal{N}(\mu_i, \sigma_i^2)$, i.e.,

$$X_i = \mu_i + \epsilon_i \quad \text{with} \quad \epsilon_i \sim \mathcal{N}(0, \sigma_i^2) \tag{5.39}$$

2. The errors ϵ_i are independent, i.e., $\mathcal{E}[\epsilon_i \epsilon_j] = 0$ if $i \neq j$.

3. The function $f(\cdot)$ is assumed to be differentiable with respect to all x_i and that the partial differential exists at $x_i = \mu_i$.

The function $f(\cdot)$ can be expanded (due to the last condition) into a Taylor series around $\mu = (\mu_1, \mu_2, \ldots, \mu_n)$:

$$f(x_1, \ldots, x_n) = \sum_{k=0}^{\infty} \frac{1}{k!} \left(\left\{ (x_1 - \mu_1)\frac{\partial}{\partial x_1} + \ldots + (x_n - \mu_n)\frac{\partial}{\partial x_n} \right\}^k f \right)(\mu_1, \ldots, \mu_n) \tag{5.40}$$

Replacing the variables x_i with the random variables X_i and substituting X_i with $\mu_i + \epsilon_i$ (where ϵ_i is a random variable) in (5.40) yields:

$$f(X_1, \ldots, X_n) = \sum_{k=0}^{\infty} \frac{1}{k!} \left(\left\{ \epsilon_1 \frac{\partial}{\partial x_1} + \ldots + \epsilon_n \frac{\partial}{\partial x_n} \right\}^k f \right)(\mu_1, \ldots, \mu_n) \tag{5.41}$$

The expectation value of (5.41) is given by

$$\mathcal{E}\left[f(X_1,\ldots,X_n)\right] =$$

$$\mathcal{E}\left[\sum_{k=0}^{\infty}\frac{1}{k!}\left(\left\{\epsilon_1\frac{\partial}{\partial x_1}+\ldots+\epsilon_n\frac{\partial}{\partial x_n}\right\}^k f\right)(\mu_1,\ldots,\mu_n)\right]$$

$$= \sum_{k=0}^{\infty}\frac{1}{k!}\mathcal{E}\left[\left(\left\{\epsilon_1\frac{\partial}{\partial x_1}+\ldots+\epsilon_n\frac{\partial}{\partial x_n}\right\}^k f\right)(\mu_1,\ldots,\mu_n)\right]$$

$$= f(\mu_1,\ldots,\mu_n) \qquad (5.42)$$

The last transformation step uses the assumption of independent errors. One can see that all terms in (5.42) involving products of two or more ϵ will vanish and $\mathcal{E}[\epsilon_i] = 0$ by definition.

The variance of (5.41) is given by

$$\mathcal{V}\left[f(X_1,\ldots,X_n)\right] = \mathcal{E}\left[f(X_1,\ldots,X_n)^2\right] - \mathcal{E}\left[f(X_1,\ldots,X_n)\right]^2$$

$$= \mathcal{E}\left[f(X_1,\ldots,X_n)^2\right] - f(\mu_1,\ldots,\mu_n)^2 \qquad (5.43)$$

Replacing the first term with the Taylor series yields:

$$\mathcal{E}\left[f(X_1,\ldots,X_n)^2\right] =$$

$$\mathcal{E}\left[\left(\sum_{k=0}^{\infty}\frac{1}{k!}\left(\left\{\epsilon_1\frac{\partial}{\partial x_1}+\ldots+\epsilon_n\frac{\partial}{\partial x_n}\right\}^k f\right)(\mu_1,\ldots,\mu_n)\right)^2\right]$$

$$(5.44)$$

Using again the independence assumption with[16]:

$$\mathcal{E}[\epsilon_i\epsilon_j] = \delta_{ij}\sigma_i\sigma_j$$

$$= \begin{cases} \sigma_i^2 & i=j \\ 0 & i\neq j \end{cases} \qquad (5.45)$$

Equation (5.44) simplifies to:

$$\mathcal{E}\left[f(X_1,\ldots,X_n)^2\right] = f(\mu_1,\ldots,\mu_n)^2 + \sigma_1^2\left(\frac{\partial}{\partial x_1}f(\mu_1,\ldots,\mu_n)\right)^2 +$$

$$\ldots + \sigma_n^2\left(\frac{\partial}{\partial x_n}f(\mu_1,\ldots,\mu_n)\right)^2 \qquad (5.46)$$

The variance is therefore given by

$$\mathcal{V}\left[f(X_1,\ldots,X_n)\right] = \sum_{i=1}^{n}\sigma_i^2\left(\frac{\partial}{\partial x_i}f(\mu_1,\ldots,\mu_n)\right)^2 \qquad (5.47)$$

---

[16] δ_{ij} is the Kronecker delta with $\delta_{ij} = 1$ if $i=j$ and 0 otherwise.

The variance of the expression is the sum of the variances of the individual errors ϵ_i weighted by the corresponding partial derivatives. A numerical evaluation of $\mathcal{V}[f(\cdot)]$ therefore needs the (partial) derivatives of $f(\cdot)$. The derivatives can easily be computed if the function f is known in an explicit form. However, in many cases f is a connection of many steps $f = f_1 \circ f_2 \circ \ldots$ and the derivatives cumbersome to compute. Automatic differentiation (e.g., using features like operator-overloading in modern computer languages or symbolic differentiation) can provide derivatives in such situations. The Gaussian error propagation model is a good choice if the computation of the derivatives is feasible, especially since it is well understood.

5.4.3 Monte Carlo methods

Monte Carlo methods (Johnson, 1987) solve problems by sampling experiments. In these experiments some or all parameters x_i of a problem as given in (5.38) are specified as random variables. The function value in (5.38) is then calculated using a set of *realizations* $\{x_i^{(1)}, \ldots, x_i^{(m)}\}$ of each random variable x_i. For large m the empirical distribution of the function value will approach its true distribution. The realization of the random variables is usually calculated based on *random numbers*. Random numbers are a sequence of digits, with the probability that – in the long run – all digits will occur equally often (Daintith & Nelson, 1991, p. 273), i.e., random numbers are realizations of a random variable with a uniform probability distribution over a certain range. Such numbers can be transformed to follow other distributions such as a normal distribution or Weibull distribution. In most cases computer generated *pseudo*-random numbers are used. Monte Carlo methods are applied not only in uncertainty propagation problems but also, for example, in numerical methods for the evaluation of multi-dimensional integrals or performance analysis of statistical methods.

The uncertainty of an expression as given by (5.38) is assessed by simulation with a series of *possible* values (Hofer & Tibken, 1992). The advantage of the Monte Carlo techniques is that the resulting uncertainty of any expressions (5.38) that can be computed with numbers x_i can be evaluated. If all x_i were random variables this would almost never be possible to be calculated analytically. Consider for example an expression given by

$$y = f(x_1, x_2) = x_1 x_2 \qquad (5.48)$$

where y is the product of x_1 and x_2. Let's call the corresponding random variables Y, X_1 and X_2. Unless X_1 and X_2 have very special distribution characteristics (e.g., both are uniformly distributed and independent) the distribution of y cannot be determined analytically. This can be easily seen by the definition of the cdf for Y:

$$\begin{aligned} P(Y \leq y) &= P(X_1 X_2 \leq y) \\ &= \int_{x_1 x_2 = y} P_{12}(X_1 \leq x_1, X_2 \leq x_2) \end{aligned} \qquad (5.49)$$

P_{12} is the joint distribution of X_1 and X_2 and is integrated along a hyperbola defined by $x_1 = \frac{y}{x_2}$. Integral (5.49) cannot be evaluated analytically in most cases. Especially if P_{12} is a multivariate-normal distribution this is not possible, even if X_1 and X_2 are independent ($\mathcal{E}[X_1 X_2] = 0$). Sometimes, however, it is possible to determine *moments* of the distribution analytically, e.g., if (5.38) is a linear expression (Kaucher & Schumacher, n.d.).

Monte Carlo techniques are not affected by this limitation based on non-existing integrals and so on. However, a sensible uncertainty estimation needs a large number of simulation runs, i.e., (5.38) is evaluated for many different sets of realizations $\{x_1^{(j)}, \ldots, x_m^{(j)}\}$. The number of necessary simulation runs is exponentially increasing with the number n of variables in (5.38) (if the random variables are not independent). The computational complexity can therefore be very high and the approach not feasible. Often, these techniques are modified using additional information gained by sensitivity analysis and other means so that only the variables making a large contribution to the resulting uncertainty are simulated, whereas others are kept fixed. As such, the dimensionality of the problem can be reduced to a reasonable amount.

Monte Carlo techniques can be used as a last resort when no other less expensive method is applicable. The implementation of Monte Carlo methods in a general framework such as VDS requires sophisticated pseudo-random number generators to be available for the data values represented. Moreover, it is necessary that *states* can be identified in the overall system in the sense, that individual simulation runs can be distinguished. This is discussed in more detail in chapter 8.

5.4.4 Other techniques

There are other techniques for the evaluation of the uncertainty of functional expression that will not be discussed in more detail here:

Analytic methods Sometimes the characteristics of the functional expression (5.38) can be examined manually with other methods such as perturbation techniques (based on the same assumptions as Gaussian error propagation) or sensibility analysis, e.g., (Mauch et al., 1992).

Non-stochastic methods Functional expressions can be evaluated using fuzzy set representations (Schmerling & Bandemer, 1985; Schmerling & Bandemer, 1990) or intervals as soon as the function $f(\cdot)$ in (5.38) can be defined for intervals or fuzzy-sets as arguments. With intervals this is very frequently used, especially to control roundoff errors as mentioned earlier.

5.5 Digital representation

The previous section has shown that physical properties affected by some uncertainty are usually characterized with a small set of real numbers, e.g., a mean value and a standard deviation. The way to describe uncertain values depends strongly on the nature of the sampling process and the phenomenon investigated. It is, therefore, impossible to define one single way to describe uncertain values applicable to all cases. A digital representation of uncertain information should meet the following requirements:

- Digital encoding means mapping to real numbers.[17] It is desirable that the representation uses a small set of numbers.

- Operators and operations for uncertain values should be defined. For instance, the arithmetic operators $\diamond \in \{+, -, \times, \div\}$ should be available along with standard functions (e.g., trigonometric functions).

- It should be possible to convert one representation to another since different data sets and their values will often use different representations.

- A suitable set of functionals (mappings to $\mathbb{R}$) should be available, e.g., $\inf(\cdot)$ (lower bound), $\sup(\cdot)$ (upper bound), $\alpha_i(\cdot)$ (i-th moment), $\beta_i(\cdot)$ (i-th central moment).

- If the representation is based on probability theory it should support *Monte Carlo simulations* (Johnson, 1987). The representation of a random variable A should be able to generate pseudo-random realizations a_i. Actually, this is just another type of functional $\phi(\cdot)$ which we call $\mathrm{rnd}(\cdot)$.[18]

One of the basic decisions when choosing a suitable model for an uncertain value is whether it can be modeled with *probabilistic concepts*. In most cases it will be the primary choice to use probability theory to describe uncertainty within scientific results. Probability theory is quite well understood and a lot of methodology has been developed. Mathematical statistics especially has benefited from this framework and has produced many useful techniques and methods for the description and analysis of data. The randomness of a property (random variable X) is defined within probability theory by a corresponding probability distribution $p(x)$. Three variants are selected here to describe a probability distribution $p(x)$:

---

[17] In fact, digital encoding means mapping to integer numbers. There are, however, means to represent a finite, countable subset of the real numbers with integer numbers (i.e., *floating point numbers*).

[18] The functional $\mathrm{rnd}(\cdot)$ actually has another "hidden" parameter (sequence number) which identifies the realization requested. During a Monte Carlo simulation run, a specific realization might be requested several times so that it is necessary that the value is always the same during one simulation step. Consider for example an expression $C = A + AB$. The simulation would calculate $c_i = \mathrm{rnd}(A) + \mathrm{rnd}(A)\mathrm{rnd}(B)$. The second $\mathrm{rnd}(A)$ must have the same value as the first one. This hidden parameter is the state mentioned in section 5.4.3.

Empirical moments: $\alpha_k = \langle \int x^k p(x)dx \rangle$ and $\beta_k = \langle \int (x - \alpha_1)^k p(x)dx \rangle$, usually α_1 (mean value) and β_2 (standard deviation).

Empirical distribution function or histogram: The distribution density $p(x)$ is described by a set of quantiles, i.e., (x_i, p_i) with $P(X \leq x_i) \approx p_i$.

Parametric distributions: A distribution type may be determined when the micro data are aggregated, i.e., the empirical distribution is approximated with a standard distribution and its parameters. Typical parametric distributions are the normal distribution (parameters μ, σ^2), uniform distribution (a, b) or Weibull[19] distribution (p, γ). Sometimes the parameters of a distribution correspond to some moments, e.g., for the normal distribution $\mu = \alpha_1$ and $\sigma^2 = \beta_2$. It is, however, different to describe a distribution solely by moments rather than by distribution type and a set of parameters.

It is not always suitable to apply probability theory to describe uncertainty. Therefore, two other non-probabilistic ways to describe uncertain values are included:

Intervals: Intervals defined by lower and upper bounds of value.

Fuzzy sets: Fuzzy sets as an extension to intervals.

While intervals are a simple yet powerful way to deal with uncertain real-valued data, fuzzy sets have not yet been used widely to describe scientific data[20] despite the attention they have received in the past few decades. The major criticism of fuzzy sets is the subjective assignment of membership degrees (definition of set membership functions) which often is not acceptable in scientific work where objectivity is an important issue. Table 5.1 summarizes the properties of these representations.

5.6 Numbers, units and dimensions

Measurement units are not directly an uncertainty issue. They are discussed here, however, in order to present a comprehensive essential model for the representation of physical quantities. Physical quantities and measurements carry in most cases a *unit*. A digital representation of such quantities preferably includes the unit of a measurement. This is particularly important in an environment addressing interoperability between various information communities. As soon as data travel across discipline boundaries the *quality* of the data is secured if values are accompanied by their units. The gain in quality is due to the possibility of performing better "type checking" when evaluating arithmetic expressions involving dimensioned numbers. Including

---
[19] Most often the special case with $p = 1$ (exponential distribution) is used.
[20] However, for some applications see (Bandemer & Näther, 1992).

Representation	parameters	inf(·) sup(·)	$\alpha_l(\cdot)$ $\beta_k(\cdot)$	rnd(·)	+, −, ×, ÷	Std.-functions
Interval	2	●	○	○	●	●
Fuzzy set	★	●	○	○	●	●
$\alpha_i, \beta_j, i \leq n, j \leq m$	$n+m$	●	★	●	★	★
n-Histogram	$3n$	●	●	●	●	●
Uniform distribution	2	●	●	●	○	○
Normal distribution	2	○	●	●	+, −	○
Other distributions	★	★	●	●	★	★

● = available, ○ = not available, ★ = variable.

Table 5.1: Properties of different representation types for uncertain values

units in a digital representation of a physical quantity means being explicit, thus avoiding misunderstandings.

Theoretically, a unit is given by the reference quantity used in a measurement. A measurement is, in principle, "finding an expression for a quantity in terms of ratio[21] of the quantity concerned to a *defined* amount of the same quantity" (Jerrard & McNeill, 1980, p. 1, emphasis added). The measurement X is identified by a tuple (x, u) of the ratio number x and the unit u with

$$X = xu \tag{5.50}$$

i.e., the tuple corresponds to a product. This means that the usual arithmetic rules can be applied for the product when performing calculations with such measurements. Consider for example two measurements $X_1 = x_1 u_1$ and $X_2 = x_2 u_2$. Using the associative law of multiplication we can deduce:

$$\begin{aligned} X_1 X_2 &= (x_1 u_1)(x_2 u_2) \\ &= (x_1 x_2)(u_1 u_2) \end{aligned} \tag{5.51}$$

The product of X_1 and X_2 can be represented as a tuple $(x_1 x_2, u_1 u_2)$ as well. This is also true for addition (and subtraction) if $u_1 = u_2 = u$:

$$\begin{aligned} X_1 + X_2 &= (x_1 u) + (x_2 u) \\ &= (x_1 + x_2) u \end{aligned} \tag{5.52}$$

---

[21] This corresponds to measurement scales identified in social sciences, e.g., (Narens, 1985; Chrisman, 1995). In natural sciences almost all measurements are of this "ratio"-type.

Therefore, X_1 and X_2 is represented by $(x_1 x_2, u)$. While this seems to be very trivial it has a large impact on the usage of such measurements. For example, X_1 and X_2 *can only be added* if $u_1 = u_2$. If the digital representation of a measurement represents X_1 and not only the dimensionless number x_1, the computing system is able to detect *incompatible* mathematical expressions.

A measurement X can be expressed with different units, i.e., $X = x_1 u_1 = x_2 u_2$. This means, that the number x_2 with unit u_2 can be calculated with a dimensionless conversion factor $f_{12} = u_1/u_2$ as $x_2 = f_{12} x_1$. Groups of units u_i with conversion factors f_{ij} between them identify a *dimension*, e.g., *length*, *time* and so on. There is usually a canonical unit u_0 which is selected according to the unit system, used, e.g., *meter* with the SI system or *centimeter* with the CGS system. The unit-conversion of ratio-type values involves only the abovementioned conversion factor. In the case of a canonical unit it is sufficient to have the conversion factors of any unit u_i to u_0, i.e., $f_{i0} := f_i$. A conversion factor f_{ij} can then be derived as $f_{ij} = f_i f_j^{-1}$. This means, that it is relatively simple to include automatic unit conversion (at least for ratio-type measurements) into a computing system.

However, full support of unit conversion and dimensional analysis needs the capability for *symbolic computation* in a manner similar to that described for the calculation of derivatives for Gaussian error propagation. The unit symbols u_i are, after all, abstract entities that cannot be handled numerically. However, for simple arithmetic expressions a dimensional analysis can also be performed without extensive symbolic manipulation capabilities with the following rules.

Addition and subtraction of two quantities $X_1 = x_1 u_1$ and $X_2 = x_2 u_2$:

1. If $u_1 = u_2$ add or subtract x_1 and x_2.

2. If $u_1 \neq u_2$ but both are elements of the same dimensional group, convert x_2 to unit u_1 by multiplication by f_1/f_2.

3. If $u_1 \neq u_2$ and neither are elements of the same dimensional group, then this is an illegal operation.

Mulitplication or division of two (or more) quantities $X_i = x_i u_i$:

1. Multiply/divide all numbers x_i.

2. Sort the units according to some scheme, transform all units of one dimensional group into their canonical unit and add or subtract the corresponding exponents of equal units. This leads to a new unit of the form $u = u_1^{p_1} u_2^{p_2} \ldots$ which forms a new dimensional group (unless such a group has already been defined for the unit u).

This means that a simple dimensional analysis can be performed without *true* symbolic computation. However, more complicated cases cannot be managed, of course. Nonetheless, a computing system providing support to tag every dimensioned number with a unit can provide more reliability when using and exchanging data as is shown in chapter 8.

5.7 Review

The objective of this chapter was to provide an essential model for the management of physical quantities in order to provide a semantically enhanced representation of measurements. This is necessary in order to provide and assess quality information when datasets are used within different information communities. One of the motivations for the VDS concept was also the embedding of methods for the calculation of derived values. If these operations happen *automatically* it is very important to guarantee a certain reliability in the result, e.g., by propagating error and uncertainty information from the source data into derived data. Also, having measurements encoded using the various methods discussed in this chapter does at least show a data user that *there is* uncertainty and that it might be a good idea to use that information. This follows, in combination with the unit inclusion, the general principle of making as much information explicit as possible.

III

Implementation

CHAPTER SIX

Case studies

6.1 Introduction

This chapter presents two interoperability case studies covering the language POSTSCRIPT and the World-Wide Web. The objective of the case studies – "success stories" both – is to show some major design decisions as a preparation for the evaluation of implementation alternatives. The studies might seem unrelated to VDS at first glance but will provide many useful insights as will be shown throughout this chapter.

6.2 POSTSCRIPT

6.2.1 Overview

The programming language POSTSCRIPT[1] was released in 1985 by Adobe Systems Incorporated (Adobe, 1985; Adobe, 1990). It is an interpretive programming language with powerful graphics capabilities which was developed mainly to describe and distribute text and graphical shapes on printed pages[2]. The motivation for the design of POSTSCRIPT was the growing number of different printing devices already discussed in section 3.5 on page 55. In particular, *raster output devices* such as laser, dot-matrix or ink-jet printers, digital photo-typesetters offered new, sophisticated features such as high-quality typefaces and powerful graphics capabilities. All these devices have in common that they produce images consisting of a rectangular array of picture elements (pixels) and that each of the pixels can be addressed individually.

---

[1] POSTSCRIPT is a trademark of Adobe Systems Incorporated.
[2] POSTSCRIPT is often called *Page-Description Language* even though this term is not appropriate for all current uses of POSTSCRIPT such as Display POSTSCRIPT.

From an application point of view, the use of raster output devices caused major problems, especially as the devices were not only powerful but also incompatible with each other. Thus, one of the design goals for POSTSCRIPT was a *standardization* of the access to these devices from the application side. An application should be able to create similar-looking output on a variety of printing devices without having to be aware of each printer's individual characteristics. This could have been realized by a software layer offering various *printer drivers* as described on page 55. The drivers would then offer a standardized interface to the application providing a set of primitives for the creation of printed output and generate the device-specific output on the application's behalf.

The design of POSTSCRIPT uses a different approach[3] which is motivated by the requirement of *distribution*. It was and is a common situation that at the time of creating print output from an application, the type of raster output device to be used for rendering is not known. In other words, it is a frequent situation that the output generation by the application and the rendering of the output on a raster output device are happening at different places and at different times. The *format* of the application's output therefore needs to be *device-independent*. Device-independence also induces resolution-independence. It is therefore clear that the format cannot be based on a *static* description of the array of pixels to be rendered on the output device. POSTSCRIPT approaches device-independence using a *dynamic* format (Adobe, 1990). A print-output or page-description is a POSTSCRIPT program which is executed (i.e., interpreted) on the target platform, i.e., the raster output device. The POSTSCRIPT program is encoded as plain-text using the popular text encoding scheme ASCII[4]. This design therefore facilitates the exchange of a page description between the composition system (data producer) and the printing system (data consumer).

Both the device-independence and the flexibility gained POSTSCRIPT a widespread popularity within its short history. At the time of writing there are more than 6 million POSTSCRIPT interpreters in use worldwide (Adobe, 1996). POSTSCRIPT interpreters are available on almost any platform, ranging from popular computing environments such as Microsoft Windows, Apple Macintosh, UNIX Workstations to specific microprocessor environments used mainly in printing devices. The POSTSCRIPT language is a full-featured programming language, with variables, control structures and many other features. Listing 6.1 shows a simple example of a POSTSCRIPT program.

---

[3] Actually, in many systems a combination of *drivers* and *device-independent formats* is used in that the drivers generate POSTSCRIPT code.

[4] However, there are also binary encoded forms of the language used within controlled environments such as Display POSTSCRIPT system. There, the communication channel between producer and consumer is transparent and can therefore be used with binary encoded forms (Adobe, 1990).

Listing 6.1 "Hello, world" in POSTSCRIPT

```
% Print the string 'Hello, world!' using a 12 point Helvetica

% move to start point
72 72 moveto

% select current font
/Helvetica findfont 12 scalefont setfont

% print it
(Hello, world!) show

% show the page (eject a paper on the printer)
showpage
```

6.2.2 Impediments

In the context of this thesis and VDS, POSTSCRIPT can be seen as a software system which is based on the exchange of *executable content*. The flexibility of the language made it a candidate for many (graphical) data exchange problems. However, there have been some lessons learned during the last decade of POSTSCRIPT usage.

Probably the largest impediment in POSTSCRIPT usage for data exchange is due to the flexibility of generic POSTSCRIPT. The content of a "POST-SCRIPT file" can, in general, only be understood using a POSTSCRIPT interpreter. Owing to POSTSCRIPT's structure, such an interpreter is a complex piece of software. POSTSCRIPT is, therefore, typically used only for the exchange between any application and raster output devices which are specific enough to be equipped with a POSTSCRIPT interpreter. However, there was and is a frequent need for various applications which run in an environment without a built-in POSTSCRIPT interpreter (e.g., on a personal computer) to identify at least some of a POSTSCRIPT file's content. We will discuss here, as a representative example, the question, how many pages are or will be produced when "printing" a POSTSCRIPT file, i.e., when interpreting the POSTSCRIPT program. Similar questions are, for example, which area of a page is covered by print output[5].

We can formulate the problem statement as follows. Given a POSTSCRIPT program (or "file" or "dataset") P, determine the number n of output pages produced by P. This is a simpler special case of the question: Split P into n

---
[5]This is needed, for example, when embedding or *encapsulating* POSTSCRIPT code into a new page description, e.g., a picture available as POSTSCRIPT code into a word processing document. The solution to this problem is also accomplished by the *document structuring convention* discussed later on.

separate POSTSCRIPT programs $P_1, \ldots, P_n$ which define the respective pages. The latter, more general problem arises very often in day-to-day computing, for example, if only a certain range of pages of a document has to be printed out or if the page sequence has to be rearranged for double-sided printing.

POSTSCRIPT defines two operators that cause the current graphics state to be printed on the raster output devices, showpage and copypage. A naive approach would scan P for each occurrence of one of these operators and report the number of found occurrences as a page count. This works for many simple cases but fails in general for many reasons, such as:

- The operator might be called within a previously defined procedure (see listing 6.2). The scanning approach only finds showpage once as it is used in the procedure definition.

- The operator might be commented out (listing 6.3).

- The operator might be executed conditionally (listing 6.4).

Listing 6.2 POSTSCRIPT program with operator in procedure, scanning counts 1 page

```
% select Helvetica 12pt as my font
/Helvetica findfont 12 scalefont setfont

/myshowthepage {
% this is my own procedure
showpage
} def

% 1st page

100 100 moveto
(first page) show
myshowthepage

% 2nd page

100 100 moveto
(second page) show
myshowthepage
```

The last point shows that it is not generally possible to determine n, i.e., the number of executions of the page output operators, *without full interpretation of the* POSTSCRIPT *program*. This means that a POSTSCRIPT interpreter is necessary to determine n. In practice, it can be implemented by adding a

Listing 6.3 POSTSCRIPT program with operator in comment, scanning counts 3 pages.

```
% select Helvetica 12pt as my font
/Helvetica findfont 12 scalefont setfont

% 1st page

100 100 moveto
(first page) show

% here we call showpage
showpage

% 2nd page

100 100 moveto
(second page) show
* and here, too
showpage
```

Listing 6.4 POSTSCRIPT program with conditional evaluation, scanning counts 3 pages.

```
% select Helvetica 12pt as my font
/Helvetica findfont 12 scalefont setfont

% 1st page

100 100 moveto
(first page) show
showpage

% 2nd page

100 100 moveto
(second page) show
showpage

% just confuse 'em
false {showpage} if
```

prologue to P which redefines `showpage` and `copypage` with new procedures which basically increment some counter and an epilogue which communicates the value of the counter to the user[6]. Moreover, P could even be so nasty that the number of pages produced is *pseudo-random* using the `rand` operator which is initialized by some external states such as the system's clock (`realtime`), the interpreter's or hardware's speed (`usertime`), or memory status (`vmstatus`), see listing 6.5.

Listing 6.5 POSTSCRIPT program randomly producing one or two pages

```
% select Helvetica 12pt as my font
/Helvetica findfont 12 scalefont setfont

% 1st page

100 100 moveto
(first page) show
showpage

% maybe a 2nd page

% seed the pseudo-random number generator
realtime srand

% get a random number in the range 0..2^31-1
% and subtract 2^30 from it. we should have about
% 50 % probability that the number is negative
rand 2 30 exp sub

% if negative show the page, else do nothing
0 le    {
          100 100 moveto
          (second page) show
          showpage
        } if
```

---

[6]Using POSTSCRIPT level 1 even this is not generally possible since a POSTSCRIPT program in POSTSCRIPT level 1 has full control over its memory (there are no local and global contexts). A POSTSCRIPT program can therefore save the memory state before calling a page output operator and then restore it again (this is a very common practice, since `showpage` re-initializes the graphics state and this might not be desired). The counter value – being a memory location within the POSTSCRIPT interpreter – will be immediately reset to its previous value after the redefined page output operator returns. POSTSCRIPT level 2 introduces the concept of global and local memory.

6.2.3 Document Structuring Conventions

The specification of the POSTSCRIPT syntax and semantics imposes no overall program structure, i.e., any sequence of conforming tokens is a valid POSTSCRIPT program. It was shown above that this leads to the necessity of fully interpreting the POSTSCRIPT code in order to assess some information about the content. It was realized early on that this is not practicable. The approach taken by Adobe was to define a kind of meta-language, called the *Document Structuring Conventions* (DSC). The basic idea is to define a standard document structure and communicate the structure with the POSTSCRIPT program with a language that is in some sense "orthogonal" to POSTSCRIPT. This means that meaningful content of DSC does not influence POSTSCRIPT and vice versa. The syntax of DSC solves this by requiring every DSC keyword to be prefixed with two '%' characters. A % character introduces a comment in POSTSCRIPT (unless it is within a string). Therefore, DSC keywords are ignored by POSTSCRIPT. DSC defines a set of keywords which both give information about the POSTSCRIPT program's content (e.g., which program created it, how many pages are in there, which fonts are used) and also identify different parts of the document (e.g., header, each individual page, trailer etc.). Using DSC a POSTSCRIPT program P can be written as

$$P = H + S_1 + \ldots + S_n + T \quad (6.1)$$

with header[7] H, single page description S_i and trailer T[8]. DSC not only specifies keywords to allow a splitting of (6.1) but also requires each S_i to be independent from others, e.g., the execution of S_i does not depend on the execution of S_j with $j < i$. This allows, for example, individual page programs P_i as discussed above to be constructed as

$$P_i = H + S_i + T \quad (6.2)$$

Listing 6.2 extended with DSC is shown in listing 6.6.

For an application program using a POSTSCRIPT program P it is very easy to extract information about P if P is conforming to DSC. There is an often used special subset of POSTSCRIPT programs, called *encapsulated* POSTSCRIPT *file* (EPSF). An EPSF program E is special in so far as it describes *only one* page and that a set of DSC keywords is mandatory. These keywords specify for example the minimum bounding rectangle on a page which is covered by E (DSC keyword %%BoundingBox) and other information which is necessary when embedding E into a document.

The example of the necessity of DSC shows a very frequent trade-off in the design of any software system. How much "freedom" is necessary? The better the flexibility the more difficult interoperability becomes.

---

[7] DSC actually identifies two separate header parts, called *prolog* and *document setup*.
[8] The "+" sign here is a sequencing operator.

Listing 6.6 POSTSCRIPT program conforming to Document Structuring Conventions

```
%!PS-Adobe-3.0
%%Title: PostScript example with DSC
%%Creator: Andrej Vckovski
%%Pages: 2
%%DocumentFonts: Helvetica
%%EndComments

%%BeginProlog

/myshowthepage {
% this is my own procedure
showpage
} def

%%EndProlog

%%BeginSetup

% select Helvetica 12pt as my font
/Helvetica findfont 12 scalefont setfont

%%EndSetup

%%Page: 1 1
100 100 moveto
(first page) show
myshowthepage

%%Page: 2 2
100 100 moveto
(second page) show
myshowthepage

%%EOF
```

6.2.4 Independence

Before discussing issues relating this case study of PostScript to design strategies for VDS, we will make one additional note, which is an issue not actually on the level of the PostScript language but which was in its design very much influenced by the flexibility of PostScript. PostScript was used very early in Apple Macintosh environments and Apple's LaserWriter was the first low-cost output device with PostScript capabilities (running version 23.0 of the PostScript interpreter on a Motorola 68000 CPU). Consequently, a driver for LaserWriter-compatible printers (i.e., PostScript printers) was an important component of the Macintosh Operating System (MacOS). The design of the driver made full use of the flexibility of PostScript and identified the printer not as a "dumb" device fed by an unidirectional communication but as a resource providing services in the sense of a distributed system. The driver engaged on a real communication with the printer, i.e., sending *and* receiving data. For example, the driver could ask the printer for the fonts available or if the Macintosh directory[9] were already available in the printer's memory. Using this information, the driver could intelligently transmit only the data *needed* to the printer and thus speed up both the communication and the processing time. Using terminology from system integration one might call this a *tight-coupling* design, i.e., the printer services are interwoven with the production system (e.g., a Macintosh personal computer).

This design was only possible because the printer's central processing unit was running a PostScript interpreter with all of its features. However, it was soon realized that the tight-coupling had some drawbacks. For instance, in a local area network with the printer connected to an intervening spooling system the printers were not directly accessible to each workstation, i.e., the workstation could not query the printer for available fonts and other state information. For a certain period the solution was to emulate a printer on the spooler. The emulation on the spooler would answer the queries for available fonts and directories, usually reporting a minimal setup and thus forcing all workstations to include all needed resources in their PostScript program. This means that – while device independence was achieved with PostScript– there was no independence from the *execution context*. This design has, however, been modified in later revisions of the Apple Macintosh printing system.

6.2.5 Lessons learned

Regarding an implementation approach, there are some relevant characteristics of the PostScript system. We will discuss them here in no particular order. It is also not our primary concern to discuss PostScript as an exchange format for *graphical* content, even though this is similar to *geographical*

---

[9]A directory is a compound data type in PostScript and is basically the only way to modularize a PostScript program.

content. The objective is rather the *identification* of some key characteristics in order to give a rationale for design decisions taken here.

Executable content A page description is a POSTSCRIPT program. The rendering happens by executing the program, i.e., interpreting the POSTSCRIPT code. This flexibility offers a wide range of possible applications.

Independence POSTSCRIPT is platform- and vendor-independent[10]. The design (e.g., using ASCII as the character encoding mechanism) is open and available to the public. Furthermore, POSTSCRIPT can also be independent from its *execution context* unless there are dependencies explicitly built in such as in the tight-coupling approach discussed above. Independence from the execution context means that there is a representation P which *fully* describes the content.

Abstraction levels The need for a DSC has shown that in systems offering such complexity as, for example, the POSTSCRIPT language itself, it is absolutely necessary to incorporate appropriate levels of abstraction with well-defined interfaces. The DSC solution can be seen as an interface definition giving a more abstract view to a document or POSTSCRIPT program, respectively. But DSC does not enforce this structuring. It is a *convention* and just meant to be good POSTSCRIPT programming style. Nonetheless, it is very common nowadays that POSTSCRIPT programs conform to DSC, showing the necessity of additional abstraction layers.

In the context of GIS one would call DSC *structured metadata*, i.e., a data set description on a high abstraction level.

Open and vendor-driven The specification of POSTSCRIPT has beend since its release, open to the public. POSTSCRIPT usage therefore boomed because documentation was available and affordable. However, unlike other approaches to openness, POSTSCRIPT is still vendor-driven and protected by copyrights.

6.3 Internet and World-Wide Web (WWW)

6.3.1 Overview

Including the Internet[11] as a case study for the design and implementation of interoperable systems is an idea which immediately suggests itself. The

---

[10] Almost vendor independent. Adobe Systems, Inc., still plays a major role in POSTSCRIPT development and holds a trademark on POSTSCRIPT. One can say that POSTSCRIPT is vendor-independent on the *technical* level and not on the *legal* level.

[11] The usual convention of capitalizing *Internet* when referring to the *connected internet* (Comer, 1991, p. 2), i.e., the large internetwork based on TCP/IP technology connecting millions of hosts worldwide, is used here.

Internet experienced enormous growth in the late eighties and early nineties and it is a showcase of a successful, large-scale, cooperative computing system consisting of many different hard- and software platforms working together. There are many textbooks explaining the services and structure of the Internet, e.g., (Comer, 1991) and its usage, e.g., (Krol, 1994). Here, we will pick only some of the design issues which helped the success of the Internet and which are of relevance for the topics discussed here:

- Interprocess communication and network-level services
- Scalability
- World-Wide Web
- Openness and standardization processes

These topics cover different areas of the Internet and its services on different abstraction levels. Some are more general design issues defining the "Internet philosophy" (e.g., scalability, openness), others are more technical in nature (e.g., interprocess communication).

6.3.2 Interprocess communication and network-level services

The development of a simple model for interprocess communication (IPC) for the 4.3 BSD UNIX operating system (Leffler *et al.*, 1989) is one of the key factors of the Internet's success. These IPC facilities were introduced into the BSD system with release 4.1c for VAX computers (Stevens, 1990, p. 261) and very soon became popular among other flavors of UNIX. They consist of an *application programming interface (API)*, called the *(Berkeley) sockets*, which offers a small and simple set of functions to implement client/server systems. The socket API contains both connection-oriented (reliable stream transport) and connection-less (i.e., message-oriented) communication, which are defined within the TCP/IP suite as TCP (*transmission control protocol*) and UDP (*user datagram protocol*), respectively (see table 6.1).

The socket interface therefore defines a simple access to the network-level services defined by TCP/IP. For many applications, it is of great advantage to use network-level services for the information exchange as opposed to direct application-level communication. Applications can use these network-level services to send information *directly* to the receiving application or the corresponding network-level services on the remote host. The applications themselves need not know how the data will be transmitted. Consider for example mail exchange. Let A be a host sending mail to another host B. A packet from A to B might travel over many routers and networks. The routing of the packets for A to B is fully under control of the network-level services, i.e., the mail sending application does not need to be aware of the routing details. A rather has the impression of talking directly to the receiving application on B. From an application point of view, the mail delivery is reliable since A knows

Function	Description	Server	Client
socket()	create endpoint	•	•
bind()	bind address	•	•
listen()	specify queue	•	
accept()	wait for connection	•	
connect()	connect to server		•
read(), write()	transfer data	•	•
send(), recv()	transfer data	•	•
sendto(), recvfrom()	transfer data	•	•
close(), shutdown()	terminate communication	•	•

Table 6.1: Berkeley sockets API

it was received by B. It is in general a fundamental difference from other inter-networking approaches that TCP/IP uses *end-to-end-acknowledgment* (Comer, 1991, p. 6). I.e., the reception of a packet traveling from A to B is not acknowledged to A by an intermediate host C but by the end-point B.

Using the socket interface, it became simple to use the system's network-level services and many programmers developed software exploiting these facilities. The simplicity of the approach – the API defines only 13 functions – led to a large available set of programs being written and new application-level protocols being defined on top of TCP/IP network-level services. Together with the widespread use of the BSD system in computer science departments at Internet-connected universities, the availability of the sockets layer considerably boosted the growth and use of TCP/IP and Internet technology.

6.3.3 Scalability

The ability of a composed system to "scale", i.e., to be more or less independent of the number of components in the system, is a very important criterion in the design of interoperable and distributed systems. Scalability means not necessarily that the system is equipped in a specific way, but that its *design* is flexible enough and does not contain fundamental barriers to later extensions. The Internet has proven that it is very scalable. Starting as a network with a few hundred nodes, it experienced an almost exponential growth with more than 10 million permanently connected nodes at the time of writing. The basic networking technology, TCP/IP, remained almost the same during 15 years of expansion. There are a few design issues which enabled the growth or at least did not limit it:

Addressing and routing Communication in the Internet or TCP/IP networks in general happens by transmitting so-called IP-packets between two nodes. Every node has a unique address which is a 32-bit integer (IP-address). The IP-packets carry both the sender's and receiver's IP-address and the payload of the transmission. There is no explicit routing information contained in the packet. This means that a system A sending information to a system B needs only to know the IP-address of B but no information on *how* system B can be reached. The determination of a communication path from A to B is called *routing*. In TCP/IP-networks, routing is established by special nodes – so-called *routers*[12] – which are connected to two or more networks. Based on routing tables the routers can determine on which of the connected networks a packet should be forwarded. The routing is based on a prefix-scheme similar to the routing in public telephone networks. Part of a host's IP-number is interpreted as a prefix[13] called the *network* address. The remaining part is the host's address on the specified network. Routing a packet between interconnected networks is only based on the prefix, i.e., the network number. A router therefore only needs to know how to reach a specific network and not every single host. As soon as a packet arrives on the network specified, it is sent to the specified host on that network.

This routing scheme is very flexible. First, it is possible to have multiple, redundant paths from A to B, enhancing the reliability of the system. Second, it is possible to change local configurations (i.e., add new hosts to a network, rearrange the local routing) without having to register this information centrally. However, the large growth has also shown some limitations. IP-numbers have to be unique throughout the entire internetwork. A centralized authority is therefore required to manage the distribution of IP-numbers. This is defused somehow by registering only network addresses (i.e., the prefix-part) and delegating the distribution of IP-numbers within the range given by the network number to local authorities. Another limitation is given by the fixed prefixing-scheme. While 32 bits allow for more than 4×10^9 distinct numbers (which is currently far enough) this is reduced by having two separate number ranges, one given by the network address and the other for the host address. Class B networks for example are in demand by larger organizations because they allow more than 256 hosts (16 bit for

---

[12]Older literature often uses the term *gateway*.

[13]There are actually two prefixes: a first two-bit prefix determines how many bits are used as a network address, i.e., 7 bits (class A network; prefix 0), 14 bits (class B; prefix 10), 21 bits (class C; prefix 110). Two other prefixes (1110 and 11110) are additionally assigned for multicasts and future use. Most implementations also allow *sub-netting* by introducing a third prefix within the host number part. Moreover, certain addresses are reserved for specific use, e.g., for broadcasts, identification of the "local" machine or multicasts.

the host address) and there are only about 16000 ($\approx 2^{14}$) class B networks available. However, many organizations with a class B network do not fully exploit the possible range of about 65000 hosts, blocking away "free" numbers.

Naming The access to resources in TCP/IP networks is based on the IP-number discussed above and an additional *port number* on each side of the communication channel. The latter is identified by the application-level protocol, e.g., HTTP (see page 134) uses port number 80. Access to IP-numbers usually takes place via a directory service that associates symbolic names with IP-numbers. The naming system commonly used in TCP/IP networks is the Domain Name System (DNS)[14]. DNS allows an n-to-n-mapping of IP-numbers to symbolic names, i.e., every IP-number may be assigned to several symbolic names (*canonical names*) and every name may be associated with several IP-numbers[15]. The naming scheme is hierarchical and allows – as with the number allocation – the management of parts (so-called *domains*) to be delegated to local authorities. The hierarchy in the symbolic names is similar to the hierarchy given by the network/host or network/subnetwork/host structure of the IP-numbers but absolutely independent of it. While the numbering is more or less an image of the physical network layout, the naming has only an organizational structure. A name can be associated with *any* number regardless of the network structure. This approach gives the system the necessary flexibility to allow operation with the huge number of hosts and users connected to the Internet. For example, local authorities may add new symbolic names as they wish or change[16] the assignment of names to numbers without having to explicitly communicate this to every connected host or user. As soon as a specific service known by a symbolic name is needed, a *name lookup* in the corresponding domain name server yields the associated IP-number. The design of the DNS as a large-scale distributed system was forced by the Internet's size and growing rate in the early eighties because it was no longer possible to manage the database of numbers and associated names in a single database (Comer, 1991, p. 8).

The same system is used in the Internet for resolution of mail addresses. The DNS is used for the management of the non-personal part in an e-mail address, i.e., the part following the "@"-character[17]. Here, the

---

[14]Sometimes also referred to as BIND, which is a specific DNS implementation.

[15]The latter is actually implementation-dependent. There are only some implementations which allow this. The benefit of having multiple IP-numbers per symbolic name is a simple method for load balancing, e.g., for heavily used WWW or FTP sites.

[16]Changes actually sometimes exhibit a latency since local name servers keep name and number tables in a cache. The entries in the cache have a specific *time-to-live (TTL)*. Within this period, a lookup will provide the cached value and not the new, changed value from the remote name server.

[17]This name space is totally different from the host name space discussed before. It is,

the development has shown that the naming system is efficient and sufficient for the delivery of mail messages but not sufficient as a general naming service. There is a need for a directory service on a higher abstraction level, i.e., associating names of persons or organizations with symbolic mail addresses. Such systems are slowly emerging and are based on X.500 or LDAP protocols (Coulouris *et al.*, 1994). The overall design of the naming system, however, was and is an important component of the Internet's scalability, mainly because name management is a distributed task.

Broadcasting Broadcasting is a form of communication that happens not between two single nodes but between one node A and an undetermined number of listening nodes. Broadcasts are typically used to look up services which are not registered in a known and accessible directory. In TCP/IP networks broadcasts are – unlike in other networking systems – used very rarely. Systems that rely on broadcasts do not scale very well, since broadcasts might exponentially flood the network in large installations. In TCP/IP broadcasts are used exclusively within single networks and are not passed over routers. For example, the physical address of a node (based on the physical communication layer) is mostly determined by sending a corresponding broadcast, e.g., using the *address resolution protocol (ARP)* in Ethernet networks. Broadcasts are usually used for services that are only locally relevant, e.g., the lookup of accessible printing devices in a local area network. Such broadcasts are implemented in TCP/IP networks in most cases on the application level, i.e., in a semantically controlled environment. Renouncing on broadcasts also has drawbacks of course. Access to services relies on knowledge of the names or addresses of the resources or appropriate directories, i.e., the absence of broadcasts needs corresponding directory services. Adding or removing resources also needs corresponding changes in the directories (e.g., domain name servers). Remedy can be achieved by self-registering systems, i.e., broadcasts are used by a service to find appropriate local directory services and to register itself with these directories.

Hardware independence The TCP/IP protocols specify a set of rules for the delivery of data (packets) over *any* physical network infrastructure. In other words, the protocols are independent of hardware design issues, i.e., network topology, speed and so on. Local area networks (LAN) within an internetwork need not be based on the same type of networking hardware. This made TCP/IP the preferred protocol suite when integrating various, previously isolated, LAN in large organizations. The

---

however, very common for organizations to use the same name for the network domain and their mail domain. The mail server for mail address user@foo.bar.com is typically located on the network foo.bar.com. Nonetheless, the latter name space is absolutely independent of the first. This is often generates confusion when setting up mail systems.

existing hardware infrastructure could be re-used with a new or additional set of protocols being transported over the physical channels.

Despite some characteristics which were also limiting the scalability of the TCP/IP technology, it is nonetheless surprising that the same technique can be used for a local area network consisting of a few hosts and for the Internet connecting, at time of writing, many millions of hosts all over the world.

6.3.4 World-Wide Web

The conception and implementation of the World-Wide Web (WWW) is certainly one of the major reasons for the increased growth rate of the Internet since 1993. WWW became a well-known – maybe even the best-known – service of the Internet. Therefore, the discussion here will be restricted to some of the basic design structures of the WWW technology, especially those with a relationship to the topic of this thesis. The WWW is based on several concepts that apply to various abstraction levels:

Hypertext Transfer Protocol HTTP (Berners-Lee et al., 1996) is a low-level component, defining an application-level protocol over TCP. The protocol is basically used to send queries to a WWW-*server* and to transmit query-results back from the server to the *client*. It is somewhat misleading that the protocol name contains "hypertext". HTTP can be and is used for many other kinds of data, and this is one of the strengths of the WWW design discussed below. The HTTP protocol uses an approach which is not difficult to implement, similar to other popular protocols used in the Internet[18]. The ease of implementation together with the free availability of specification and sample implementations quickly led to many HTTP-clients (browsers) and HTTP-servers for most platforms and operating systems.

Hypertext Markup Language HTML is a simple markup language used to create hypertext documents that are platform-independent. HTML documents are *Standard Generalized Markup Language (SGML)* documents that are appropriate for representing information from a wide range of domains (Berners-Lee & Connolly, 1995). HTML defines a set of markup elements through the HTML *Document Type Definition* as a formal SGML syntax definition. There are several characteristics that contributed to the success of the WWW.

First, it is possible to render an appealing visual representation using the markup information in an HTML document. HTML defines structure

---

[18]Simplicity means that the protocol is *state-less*, i.e., it is not necessary for the server or the client to implement a state-machine. Moreover, the protocol is based on an ASCII-encoding, i.e., a human-readable form similar to most popular Internet application-level protocols such as FTP, SMTP, NNTP and so on. Such an encoding is much easier to implement and to debug than a binary encoding.

elements such as headings, bulleted or numbered lists, emphasized text. Furthermore, it allows embedding of pictures and other non-text information. HTML allowed for the first time information to be presented on the Internet in a graphically sophisticated way.

Second, it is possible to embed so-called *hyperlinks* in an HTML document which define relationships between a text fragment (e.g., some words or a picture) and another document. The important innovation to earlier approaches such as *Gopher* (Anklesaria et al., 1993) was that the "other document" might be any resource that can be defined by a URL (see below), e.g., an HTML document available on a *remote* HTTP-server, a document available on an FTP-server or even a person reachable via e-mail. This concept allows the creation of information networks with any associations between single information items. The WWW consequently is a single gigantic network of information items.

Uniform Resource Locators A URL is a compact string representation for a resource available via the Internet with a well defined syntax and semantics (Berners-Lee *et al.*, 1994). The main objective of the URL specification is to define a *unified* name of a resource on the Internet, independent of the access method. A URL consists of a part identifying the *access scheme*, i.e., the method being used to access the resource, and a part containing *scheme-specific* information, i.e., a description of the resource pertinent to the access scheme. Table 6.2 shows some examples of URLs for various access schemes.

The use of a URL as an identification mechanism in hyperlinks made the WWW a very successful *integration technology*. Using agents (browsers) that were capable of dealing with various application-level services such as HTTP, electronic mail or FTP, Internet users were able to access the resources from a single and user-friendly application. In pre-WWW times, each application-service on the Internet was typically accessed with specific applications which were mostly command-line tools for initiated users only.

Based on these concepts, the WWW can be seen as a *distributed* and *interoperable* information system: distributed, because it is explicitly based on associations of resources available on different computers and interoperable, because it supports various application-level services, and because both user agents and servers conforming to the standard are available for most platforms and operating systems.

We will discuss here a last, often overlooked feature of the WWW system which is of particular interest to this thesis. The concept is called *content negotiation* and is important because it separates a resource from its *representation*. Content negotiation is based on a typing mechanism according to the *Multipurpose Internet Mail Extensions* (MIME; (Borenstein & Freed, 1993)) which assign every message delivered by a server a specific *content type*.

Service	Access scheme[a]	Scheme specific[b]
World-Wide Web	http	//host/path
		http://www.foo.bar/sample.html
File-transfer	ftp	//host/path
		ftp://ftp.foo.bar/sample.dat
USENET news	news	//host/newsgroup/article-number
		news://news/comp.foo.bar/82763
Electronic mail	mailto	mail-address
		mailto:foo@bar.com

[a] There are more access schemes specified in (Berners-Lee et al., 1994) than these examples here.

[b] The specification contains additional items such as port numbers (for hosts that are not using the access scheme's default port), user names, passwords and so on.

Table 6.2: URL examples for various access schemes

The HTTP specification *requires* a server to specify the content type of a reply. The content type is identified by a *type* and a *subtype*. The type identifies a general categorization of the information, e.g., text, audio, video, image and so on. The subtype specifies a specific format, i.e., a digital representation. Table 6.3 shows some examples of MIME content types.

The interesting part now is that HTTP also defines that a client can send a *list of accepted types* with a request[19]. This means that a specified URL or its *path*-part need not be interpreted as a fixed resource identification but as a more abstract description of a resource which might be available in various forms. The HTTP server can – based on the accept list sent by the client – decide, which of the available representations best match the client's requirements and send an *appropriate* representation. Consider for example, that a client requests an URL http://foo.bar/graph, i.e., it requests /graph from host foo.bar using protocol HTTP. Let's assume that the client also sends an accept list containing the MIME types text/plain, text/html and image/jpg. An intelligent HTTP server could now determine the best representation available for the path /graph and return the best type match to the client. If there were two representations – /graph.gif as GIF-file and /graph.jpg as JPEG-file – available on the server it would return /graph.jpg, because its type matches the client's accept list. This technique

---

[19] Actually, it is considered good practice to do so even though it is not formally required.

Type/subtype	Description
`text/plain`	Plain text using the US-ASCII character set
`text/html`	Text encoded as HTML document
`image/gif`	Image in GIF-format
`image/jpeg`	Image in JPEG-format
`application/pdf`	Adobe PDF document
`application/x-tar`	UNIX tar-file
`audio/x-wav`	Sound clip in WAV format

Table 6.3: Examples of MIME content types

is particularly useful for multimedia data types such as images, audio-clips and video-sequences. A server can store (or dynamically derive as a VDS!) various representations of an object and deliver the appropriate types depending on each client's accept list[20]. Another useful application of content-negotiation is within multilingual environments. In addition to the data type, a client can also send the preferred *language*[21] in a request. Based on this information, a server can determine the best matching language version of a document.

This technique of managing and accessing *multiple representations* is a fundamental component of the WWW system. Unfortunately, at the time of writing there are not yet many implementations of useful content negotiation schemes. This is partly due to the possibility of specifying wild cards within a content type. A client may also send an accept list containing `image/*` or even `*/*`. The first would accept all image subtypes while the latter accepts *everything*. Many browsers send the latter *catch-all*-type as the last element in their accept list, i.e, they "want to have it, whatever it is". Moreover, typical clients do not include priorities for the various types so that it is almost impossible for a server to determine an appropriate type. It is also additional work for the data producer (content provider) to provide multiple representations of an object, since there are almost no tools that are able to

---

[20] In actual implementations, there are currently three approaches for type selection. The first uses a so-called variants file which contains references to all variants of a specific path. The second uses a filename extension mechanism which identifies multiple representations with various filename extensions to a unique root name. The last approach uses the MIME type as an additional attribute of the object. This approach is only possible if the storage system for the object supports additional attributes. Examples are database management systems or file systems with strong file typing such as in the Apple MacOS.

[21] This is not given by a MIME type but a simple ISO language code.

provide *dynamically derived* representations. A data producer has therefore to provide static versions of all variants of an object. While language versions have to be manually translated anyhow, it would be easy to provide dynamic generation of various representations of multimedia data types.

6.3.5 Openness and standardization processes

The TCP/IP technology did not arise from a specific vendor or standardization society such as the International Standardization Organization (ISO). All specifications are created and reviewed by volunteers within the so-called *Internet Architecture Board (IAB)*. This organization contains two groups: the *Internet Research Task Force (IRTF)* and the *Internet Engineering Task Force (IETF)*[22]. Both divisions contain a number of working groups dedicated to specific areas of interest. The work of these groups and other volunteering individuals is documented in a series of technical reports called *Request For Comments (RFC)*. At the time of writing, there are almost 2000 RFCs available, covering a broad range of topics such as IAB organization, low-level protocol specifications, bibliographies and so on. The RFCs are maintained by a group called *Internet Network Information Center (InterNIC)* and available to the public. There is no proprietary or hidden standardization within TCP/IP. Vendors compete by offering better implementations of these open standards instead of hiding such information. It was surprising to many people that an open, somewhat anarchic organization of the standardization process was able to produce a technology that became the *de-facto* standard for internetworking, despite similar attempts in the industry. The key to the success was and is certainly the *openness* of the entire system. This did not only enable many people to use the technology, but also encouraged many to enhance it, provide new application-level services, and improve network-level services.

6.3.6 The future: IPv6

The future development of TCP/IP is another interesting point in studying the characteristics of the Internet, especially because it is influenced by its own history. The reasons for changing the basic TCP/IP technology, which have remained almost unchanged since its inception in the late seventies, are (Comer, 1995):

- New computer and communication techniques such as wireless networks and point-to-point satellite communication offer new features that can be exploited.

- New types of applications and application architectures are becoming popular, needing specific networking infrastructures. Video communi-

---

[22]This is the structure after the reorganization in 1989 (Comer, 1995). The splitting into a *research* and an *engineering* part became necessary due to the large size of the Internet on which millions of people depend for daily business.

cation for example needs resource pre-allocation on the network in order to guarantee a sufficient network bandwidth.

- The growing number of users on the Internet and complexity of the applications led to a dramatic increase of network traffic load. This increase is even higher than the increase in the number of users because the average amount of data transferred per user is increasing as well due to multimedia-based systems such as the World-Wide Web.

- The Internet is used all over the world and cannot be managed by centralized authorities any more. Different countries might use varying policies for Internet operation. It is therefore necessary to find ways to accommodate new administrative authorities.

A major impact will be exerted by the new version of the *Internet Protocol (IP)*, the basic protocol used for addressing and routing in the Internet. Currently, version 4 (IPv4) is used which has some drawbacks as discussed on page 131. A new version, called "IP – the next generation" (IPng) or IPv6 introduces some fundamental changes, including (Comer, 1995, p. 493):

Larger addresses IPv6 defines an address space of 128 bits, four-times as long as the 32 bits in IPv4.

Support for resource allocation IPv6 allows pre-allocation of network resource to allow communications with a guaranteed bandwidth and delay.

Protocol extensions IPv6 allows extensions to itself. The extension mechanism is pre-defined, so that new hardware or applications can be supported by IPv6 without the need to have a new protocol revision or a specialized application-level service.

The last point shows perhaps the most significant change because it specifies not only a *norm* but also the *procedure to change the norm*. In politics, it has been known for a long time that a constitution should contain rules for change or amendments.

6.3.7 Lessons learned

The Internet has introduced a vast number of applications so that almost *everything* can be proven with an example from the Internet, that is, the size of the Internet enables an example to be found for almost *every* proposition in there. However, there are a few key characteristics of the Internet that should be kept in mind when implementing VDS and related systems:

Flexibility The discussion of content negotiation as an example of multiple views within the Internet or the protocol extension mechanism of IPv6 show the importance of flexibility of applications in an interoperable environment. This may be a trivial statement, but it is still very important. Unlike closed systems where communication partners are very well informed about their counterpart (e.g., data types and protocols supported by the counterpart), exhaustive knowledge of communication partners cannot be expected in large interoperable systems. Therefore, it is, for example, necessary to support appropriate mechanisms for negotiating and querying capabilities.

Vendor and platform independence As with the POSTSCRIPT case study above, platform- and vendor-independence or the availability of implementations on most platforms is a major criterion for the Internet's success. An implementation must therefore be aware of any steps introducing major implicit or explicit dependencies on a specific platform or vendor's product.

Openness The independence discussed above is clearly also a result of the openness of the specification. Moreover, the Internet did somehow introduce a "philosophy of openness" into the computing community. As a result of that, it will become increasingly more difficult for vendors and organizations to promote technologies that are based on proprietary standards and closed systems.

Simplicity Most of the application-level protocols are very simple. For example, all major protocols such as FTP (file transfer), SMTP (mail exchange), NNTP (Usenet-news exchange) and HTTP are based on three- or four-letter commands exchanged between client and server. Interprocess networking is another example of a simple approach to a possibly complex problem. Most concepts within the Internet are designed in accordance with the "keep it small and simple" rule.

Scalability The scalability as discussed above is certainly an important characteristics. Design decisions should always be checked against their scalability.

6.4 Review

POSTSCRIPT and the Internet are two examples of systems that were designed to be interoperable and that were *successful* because they were interoperable. The case studies have shown that in an interoperable system, it is important that there is something on which all cooperating parties agree. This implies that there must be something that can be expected from a cooperating party. In POSTSCRIPT, this is the programming language POSTSCRIPT with its syntax and semantics, and the availability of an *interpreter* for that language. In the Internet, these are agreed protocols and interfaces.

The simpler the part that cooperating parties have to agree on, the higher the acceptance will be. Consider as an non-computing example various transportation systems. Railroads, for example, depend on specifications for parameters such as the gauge of the tracks and the voltage of the electricity. Interoperability of different railroad networks is given if the networks use the same specification or if special equipment is used which allows dynamic adaption to different gauges or voltages (e.g., a "VDS-train"). Interoperability on road networks is typically much easier to achieve because the specification of the interface (a road in this case) is very simple and defines that it has to be more or less flat. The more specialized interface of railroads (e.g., *exact* gauge and voltage) allows one to build more efficient transportation systems in terms of transportation capacity; whereas the low-profile interface specification of roads (e.g., flat and not too narrow) allows basically *every* automobile to function on *any* road.

The POSTSCRIPT example shows this trade-off very clearly. On the one hand it makes sense to invent some specification which enables, "semantic heterogeneities" such as differences of printing devices to be overcome and a high degree of flexibility and expressivity to be provided. This is given in POSTSCRIPT by using a dynamic representation, that is, a description using a programming language. On the other hand, this might require some complex system to make use of it, such as a POSTSCRIPT interpreter or railroad tracks, something impeding interoperability. The Internet compares with the automobile case. The agreed protocols are simple and can be implemented on almost every platform, and as such, enable a high degree of interoperability. One the other hand, the TCP/IP protocols are not the fastest network protocols known.

The next chapter discusses implementation alternatives which can be used for the implementation of such interoperable systems and which try to provide both high expressivity and ease of interoperability. The former is given by using object-oriented concepts, the latter by using distinct interfaces.

CHAPTER SEVEN

Strategies

7.1 Introduction

Both the global characteristics (effects of the discretization of the domain $\mathbb{D}$) and the local characteristics of continuous fields (representation of individual values from the range $\mathbb{V}$) were discussed in previous chapters as a theoretical background for digital representation of continuous fields. This chapter will now step back from the example of continuous fields to a more general view of spatial information and explore possible implementation strategies for the concept of virtual data sets introduced in 3.5.

VDS was introduced as an abstraction from concrete representation techniques by defining access to the data through well-defined interfaces or messages in an object-oriented model. The abstraction – hiding low-level details on data models used and encapsulating data set "internals" – is one of the core components of the approach. The motivation to introduce an additional abstraction level was an improvement for the *distribution* of spatial information. An implementation strategy therefore has to address the distribution of VDS as a major requirement.

There are many alternatives for the implementation of distributed systems in general and distributed *object* systems (as required by the VDS concept) in particular (for an overview see (Valdés, 1994) or (Adler, 1995)). Experience with large, monolithic applications in the past few decades has produced numerous concepts both in research and industry. This chapter discusses three implementation alternatives covering Microsoft's Component Object Model (COM), Object Management Group's (OMG's) Common Object Request Broker Architecture (CORBA) and an approach based on the Hypertext Transfer Protocol (HTTP). Finally, the software system Java will be presented which has been chosen as an implementation environment here.

As discussed previously, the main objective is to create an abstraction layer which allows access to geodata through a well-defined interface. Moreover, the implementation of the interface is intended to be *specific* to a dataset. The interface defines a set of *accessors*[1] to the dataset, i.e., methods that provide a view of the model given with a dataset. This means, that the *implementation of the interface* and the *state* of the dataset may not be separated. It suggests the use of a distributed *object* system which associates an interface with an object.

The distributed object systems discussed have both been available for several years and been used in many applications. They are often used in three-tier architectures as is shown in figure 3.4 on page 59. The following discussion is intended to show the major principles in order to analyze their feasibility for a possible implementation of VDS.

7.2 Component Object Model

7.2.1 Overview

The *Component Object Model (COM)* (Stafford & Powell, 1995) is a core part of Microsoft's OLE 2 technology[2]. The COM consists on the one hand of a *specification* that defines a binary standard for object implementation that is independent of the programming language used. This binary standard allows applications to communicate through object-oriented interfaces without requiring either to know anything about the other's implementation (Brockschmidt, 1994, p. 5). On the other hand COM also provides an *implementation* which consists of a set of fundamental functions that are used to instantiate so-called *component objects*. These functions can create instances of such objects based on a central system-wide registry of available classes. For example, a class registered might be "word processor document". The registry knows that the respective implementation of the class's interface is given within a specific application. Whenever a "word processor document" needs to be instantiated functions provided by COM can locate the implementation and provide the caller with an access path (e.g., a pointer to the function table of an interface) to the object created. The object can be manipulated and queried through the access path given without any user interaction in the application containing the object's implementation. This allows, for example, another application to create and embed a "word processor document" within its own execution context, using the functionality provided by the word processor application.

---

[1] *Accessor* is defined as "any relatively standard small and simple method that is used to either get or set the value of an instance attribute" (Firesmith & Eykholt, 1995, p. 10).

[2] OLE stands historically for *Object Linking and Embedding*. However, OLE 2 is a technology that goes far beyond earlier versions. Object linking and embedding refers to a small subset of what is specified within OLE 2 and deals with storage and editing documents of an application within documents created and managed by another application.

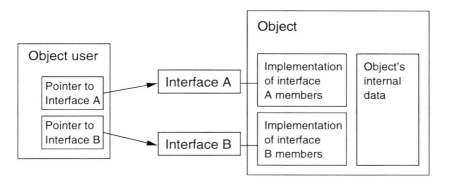

Figure 7.1: Interfaces in the component object model (Brockschmidt, 1994, p. 11)

7.2.2 Interfaces

COM is based on the notion of *interfaces*. An interface in COM terminology is a collection of function definitions (i.e., signatures[3]). An instantiated component object provides the *implementation* of some interfaces. Object users can access an object's internal state only by means of the functions (or methods) given in the interfaces that are implemented by the object. Figure 7.1 shows the call of object methods through the interfaces.

COM offers a certain flexibility for the access through interfaces in that an object can be *queried* for the interfaces supported by the object. This is achieved by specifying that every object *must* implement a base interface called IUnknown, providing a small set of methods that can be used to enumerate available interfaces and handle basic lifetime information, e.g., implementing a reference counter for object users. All interfaces are derived from IUnknown and provide its basic methods. An object user uses the QueryInterface function call within IUnknown to retrieve pointers to all interfaces implemented by an object. The ability to expose various interfaces is very useful for multipurpose objects. Consider for example an object which might be "printed" (i.e., the object knows how to produce a visual and printable representation of its state) and which might be "stored" on secondary storage (i.e., the object knows how to save and restore itself, for example, to and from a stream). The object then might implement interfaces such as IPrintable and IStoreable. These interfaces contain the methods used for printing (e.g., PrepareForPrinting(), PaintPage(), and so on) and persistence (e.g., SaveToStream(), LoadFromStream()). An object user interested only in loading and storing objects needs only to be aware of the IStoreable interface of the object, whereas another object user might only be

---

[3] A formal declaration of the function by its name and types of return value and parameters.

interested in the object's printing capabilities given through IPrintable. Using the QueryInterface function call, an object user can determine whether the given object implements, for example, IPrintable. Every interface is equipped with a unique *interface identifier (IID)* which is used to identify and distinguish interfaces. This implies that every distinct interface possesses a unique IID within the totality of interoperating applications and that all interoperating applications know the specification of the interfaces given by an IID. The uniqueness can be simply achieved using a powerful pseudo-random number generator and information such as a host's Ethernet address which is unique worldwide. IID are 128 bit numbers and the chance that two randomly generated numbers using additional location-specific information is very, very small. However, the second implication of identification by a number has major consequences. An object user can only use an object through a specific interface if it knows the *specification* of that interface, e.g., which functions are defined within the interface and what their parameters are. There is no mechanism to query an interface for the signatures of the functions defined therein. In other words, it is not directly possible to query an object for a contract to use it. The contract needs to be negotiated *beforehand*, for example by distributing a C/C++ include file by the interface creator. This include file would then be used by an object user to create an application containing the necessary abstractions (function prototypes) to call the interface's functions.

7.2.3 Feasibility

The COM is a technology which has – as core part of OLE 2 – found widespread use at the time of writing due to the large market share which operating systems and applications from Microsoft have. From this point of view it is sensible to use COM as a basis for VDS and an implementation of OGIS in general. There are, however, some drawbacks which lead to other choices as an implementation base:

Platform dependence The COM technology was designed and developed for use within the family of operating systems produced by Microsoft. There are efforts to provide libraries and tools on other platforms, but COM use today is almost exclusively within Windows-based systems. When used locally, COM also enforces that the object accessed is written for the same architecture as the local machine.

No distribution The COM specifies the access to objects whose implementation is available *on the local machine*, either within a (dynamically linked) library or a stand-alone application. It does not contain any aspects of distribution. However, there are concepts providing some starting points for a possible distribution within COM such as the so-called *lightweight*[4] *remote procedure call (LRPC)*. This mechanism is

---

[4]Lightweight, because "remote" means "other process context" and not "other machines on the network".

used to call functions implemented in a stand-alone application and running within an own execution context. The COM libraries provide, for example, the necessary functions for *marshaling*[5] and *un-marshaling* function arguments and results within a LRPC.

Microsoft recently extended COM to *Distributed* Component Object Model (DCOM) and submitted its specification as an Internet draft. DCOM is based on the Open Software Foundation's DCE/RPC specification (DCE/RPC, 1994) and is also planned to be supported by CORBA (see below). However, at the time of writing, it is not at all clear where this technology will be heading[6].

Interface specification and semantics Within COM there is no exactly-specified way to exchange interface specification and semantics between object users and object implementors. An object implementor must be able to inform the community of object users about the specification (what functions are available) and semantics (what the functions are supposed to do) of new interfaces. Because component objects *do not* carry this information by default, it is necessary to provide it by other means. If someone defines a new interface he or she would probably distribute the IID of the interface together with include files defining the interface's functions for access from C or C++ programs and text documentation about the interface's semantics. Developers using other programming languages would possibly transfer the C/C++-include files into their preferred representation using the documentation that has (it is hoped) been provided.

Complexity The overall COM, including upper-level components of OLE 2 such as *compound documents*, *automation* and so on form a complex software system which is accessed by an application programming interface of more than 100 functions (Brockschmidt, 1994, p. 5). Its design is guided by exhaustiveness and richness of features and not by simplicity and complexity reduction.

Due to these reasons, COM was not chosen as an implementation architecture for further investigation in the first place. Note, however, that COM has no fundamental drawbacks concerning an implementation of VDS. A slightly different weighing of the various factors could have resulted in another choice.

---

[5]To put it simply, marshaling means conversion into a process-independent canonical form. Sometimes this also covers resolving byte-ordering issues.

[6]The DCOM specification has been submitted as an Internet-draft. These are working documents of the IETF and its working groups that will be submitted eventually as an RFC. According to the IETF policy, Internet-drafts should not be cited or quoted in formal documents since they are not an archival document series. The DCOM specification is therefore not referenced here.

7.3 Common Object Request Broker Architecture

7.3.1 Overview

The *Common Object Request Broker Architecture (CORBA)* is a specification of a distributed object system created by the *Object Management Group (OMG)*, a consortium of most major soft- and hardware vendors. The overall objective of CORBA is to provide a framework for interoperating applications using an object-oriented model. In that sense, CORBA merges *object computing* and *distributed computing* into *distributed object computing* (Otte et al., 1996, pp. 1–4). The core part of the specification is the definition of common interfaces and services of *Object Request Brokers (ORB)* that applications use as an underlying communication mechanism. The ORB is a central part within the overall OMG Object Management Architecture which additionally specifies *object services* and *common facilities*. Compared with traditional environments for distributed computing such as remote procedure calls or interprocess communication as discussed on page 129, CORBA enhances the computing model in several ways:

Flexible client/server model CORBA allows various interaction models between applications to be adopted. Besides the classical client/server model, where typically many *clients* issue requests to one identified *server*, CORBA enables also direct peer-to-peer models. CORBA only cares about where the implementation of a specific object resides. Within an application using many objects, the implementation of the objects might be available on various nodes within a network. The application could contain the implementation for some objects itself and use some other objects with an implementation on a remote host. However, the implementation of all methods defined within one interface is located on the same node.

Broker as intermediate layer The CORBA computing model is based on a kind of *software bus architecture* in the sense that the broker represents an intermediary between clients and servers. All requests to a server are managed by the broker. This adds a substantial level of flexibility for the location and allocation of services, using multiple servers serving multiple clients.

Flexible server architecture Based on the intermediate layer, a server can be realized in several ways. Most importantly, a server need not be a *single* process as is often assumed in classical client/server computing. This is particularly interesting for systems expecting heavy workloads because many separate processes (and processors) can be allocated to serve client requests.

Asynchronous and synchronous communication CORBA allows asynchronous and synchronous communication between clients and servers.

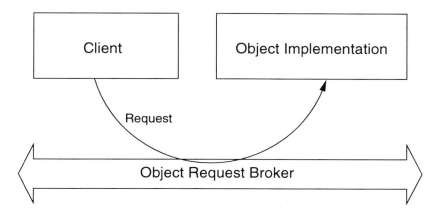

Figure 7.2: Basic architecture of CORBA

Asynchronous communication is often avoided in classical client/server computing due to the additional complexity introduced by multi-state clients and servers. In many cases, however, asynchronous communication might be the preferred choice in order to avoid blocking applications when issuing lengthy requests.

Object model Probably the most fundamental difference from classical approaches is the adoption of an object model in CORBA. The object model builds on familiar object-oriented concepts such as *abstraction, encapsulation, inheritance* and *polymorphism*. The objects in CORBA's object model provide a set of services that can be requested by a client. These services are implemented on the server and called using an *object reference*, i.e., a request is always paired with a *target object*. That is, the object model is visible on both the client and the server sides in that the client needs to specify the target object when requesting a service and a server is informed about the target object receiving the request[7].

Figure 7.2 shows the basic architecture of CORBA. A client invokes a request on an object through the ORB. The ORB forwards the request to the object implementation. Depending on the communication model, the implementation will return the results of the request on completion (synchronous) or immediately return for a deferred execution (asynchronous). In the latter case, the client has to track the execution state and eventually retrieve the request's results using the respective mechanisms provided by the ORB.

---

[7]A request can additionally contain parameters and an optional *context object*.

7.3.2 Interfaces

CORBA, similarly to COM, is based on the definition of interfaces that client objects call and object implementations provide. For that purpose, CORBA specifies a language which completely defines the interface and fully specifies each operation's parameters. This language is called *Interface Definition Language (IDL)* and is defined in the specification document (CORBA, 1992). Its syntax resembles C++, but there are some fundamental differences:

- IDL is only used to describe interfaces. It therefore lacks any constructs that are useful for an implementation. There are no "statements" in IDL. Compare also with the EXPRESS data definition language discussed in section 3.3.1 on page 40.

- Parameters within the declaration of an operation contain not only a parameter's type but also an additional directional attribute informing both the client and the server of the direction in which the parameter is to be passed.

- An optional attribute in the specification of an operation indicates which expressions might be raised as a result of an invocation, i.e., the possible exceptions are known in advance. This is sometimes called *declarative exception handling*.

- Defining an attribute in IDL ("member variable" in C++) actually only provides a pair of accessor functions[8]. IDL has no concept of "storage" and every access to an object's state is through an operation.

Listing 7.1 shows a simple example of an interface declared with IDL.

Using interface definitions provided as IDL code client and server developers can independently write code to invoke and execute requests, respectively. Most vendors of CORBA implementations provide *IDL compilers* that can be used to generate the appropriate client stubs[9] and server skeletons. Such an IDL compiler or its products, respectively, are language dependent. Figure 7.3 shows the information flow in this process.

7.3.3 Feasibility

CORBA's approach to distributed object computing looks very appealing for an implementation of VDS. It offers a clear scheme for *interfacing* using IDL. It provides *encapsulation* by separating method execution and method invocation. And, it is built around an *object model*, making the transition from the object-oriented concept for VDS to an object-oriented implementation straightforward. However, there are a few drawbacks which finally led to another choice as we will see later on.

---

[8] See footnote 1 on page 144.
[9] A local procedure corresponding to a single operation which invokes that operation when called (Firesmith & Eykholt, 1995, p. 430).

Listing 7.1 Example of CORBA interface definition language

```
module shapes {

  struct Point {
    int x;
    int y;
  };

  // declaration of exceptions OUT_OF_BOUNDS and NEGATIVE_VALUE
  // needed here

  interface shape {
    void moveto(
      in Point newpoint // moves shape to new point newpoint
    ) raises (OUT_OF_BOUNDS) ;
    void scale(
      in float factor   // scales shape by factor factor
    ) raises (OUT_OF_BOUNDS,NEGATIVE_VALUE);
    void boundingbox(
      out int x,        // lower left corner of MBR
      out int y,
      out int w,
      out int h         // width and height of MBR
    );
    float area();       // calculate the shape's area
  };

  interface rectangle: shape {
  // inherits from shape
  // implementation must provide
  // operations declared for shape
  // ...
  };

  interface circle: shape {
  // inherits from shape
  // implementation must provide
  // operations declared for shape
  // ...
  };

  interface roundedrectangle: rectangle {
  // inherits from rectangle
  // ...
  };
};
```

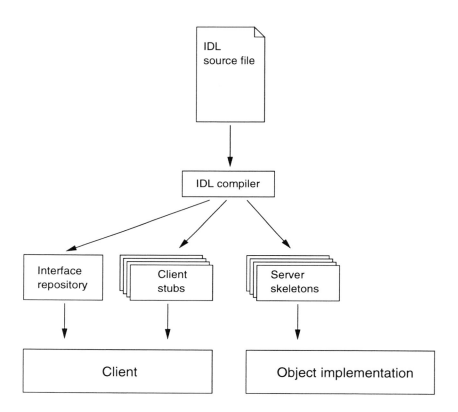

Figure 7.3: IDL compiler, client stub and server skeleton

Vendor dependence Similarly to COM, where platform dependence was a major criticism, CORBA exhibits a certain vendor dependence. Up to version 1.2 of the specification many implementation-level details were left open to the vendor of CORBA middle-ware. Consequently, it was easy to achieve interoperability between applications written on top of one vendor's CORBA implementation, whereas interoperability between different CORBA implementations was non-existent. The vendor dependence concerns both the language binding (e.g., the type of code produced by the IDL compiler) and the communication details within the software bus (ORB). This major weakness was realized within OMG and version 2.0 of the specification (CORBA2, 1995) contains the definition of a protocol that can be used to interconnect ORBs from various vendors over TCP/IP networks. This protocol is called *Internet Inter ORB Protocol (IIOP)* and addresses the interoperability issues mentioned before[10].

Missing object reflection An object user usually identifies the operations that can be invoked on an object using the interface specification available as IDL code. Vendor-specific IDL compilers support the identification by producing the appropriate client stubs from a given IDL file. In systems as proposed by VDS, it is, however, necessary for a client to be able to determine some of an object's characteristics at runtime. Or, to put it simply, it should be possible to use an object and its methods even if the object was not known at compile time by the object user. CORBA foresees such object use by virtue of the *dynamic invocation interface*. This interface allows the dynamic construction of object invocations, i.e., invoking an operation by specification of the operation by name and the corresponding parameters. Information about the operations available in an interface, their signatures and so on, is provided by an *interface repository*. There is a set of pre-defined operations for the access of the interface repository. The detailed implementation, however, is left open to the vendor and the programming language used: "The nature of the dynamic invocation interface may vary substantially from one programming language mapping to another" (CORBA, 1992, p. 35). This also applies to the interface repository: "The features and tools each vendor provides for using the interface repository might be important [...] in selecting a CORBA vendor." (Otte *et al.*, 1996, pp. 3–13).

Object migration CORBA itself does not specifically address *object migration* issues. It is left open to the CORBA vendor how exactly a server containing an object implementation is to be located and addressed. In most current implementations, there is typically a number of nodes in a network serving operations (object implementations) and other nodes invoking the operations (clients). Being a *distributed system*, the clients

---

[10] However, at the time of writing, there was no useful IIOP implementation available.

and servers are assumed to run on different hosts, or at least, different processes or address spaces, respectively. VDS requires objects *and their* implementation to migrate from data producers to data users. It is not sensible to require all operations to be executed on a data producer's host since it cannot be guaranteed that this service is always available and available with a sufficient bandwidth. A *local execution* model, where the object's implementation is transferred/migrated to the client and executed in the client's context, is therefore required for many operations. These operations can be characterized as methods that are small and independent on external resources. Examples are conversion routines or local interpolation on continuous fields.

Despite these drawbacks CORBA is a promising technology for the implementation of distributed systems. As soon as vendor independence and large-scale distribution[11] are achieved, as planned with IIOP, it will possibly gain widespread use. Simultaneously ongoing standardization attempts for the access of object-oriented database management systems of the *Object Database Management Group (ODMG)* (Cattell, 1994) can support the acceptance of CORBA. Interoperable access layers to database management systems (e.g., Open Database Connectivity (ODBC) (ODBC, 1992)) are, after all, currently the most widely used type of middle-ware and, consequently, important for the success of CORBA.

Another circumstance possibly supporting the success of CORBA is the standardization procedure adopted. The CORBA standardization process is, unlike COM, driven by a consortium of (roughly) equal members. While a single-vendor standardization might lead to quick time-to-market and a general market dominance, as has been seen in the last decade by Microsoft's operating systems and also the POSTSCRIPT example discussed in section 6.2, large-scale acceptance over platform and vendor barriers such as with TCP/IP is only achieved by an open, "democratic" approach[12]. This establishes a certain optimism for CORBA's future. In order to compete with other approaches for distributed systems, however, CORBA vendors need to provide tools which ease the development of CORBA-based systems. Currently, it is not at all a routine task to implement a distributed application on top of CORBA. There is a certain imbalance in the three approaches discussed in this chapter in that there are already some implementations which allow, for example, the use of COM or Java over CORBA. The strategies presented are not only implementation *alternatives* but also implementation *aspects*.

---

[11] I.e., wide-area or global-area networks. This raises, of course, all the questions of scalability asked before in the context of the Internet. CORBA implementations have not yet been proven to scale by several orders of magnitude. There are, however, no fundamental impediments that seem obviously to hinder large scalability.

[12] The Internet's standardization procedure is sometimes also called "anarchic". Stressing political comparison it might rather be called a "grass-roots movement". In this comparison, CORBA might be seen as an "oligarchy" as compared with the "dictatorship" of Microsoft.

7.4 Other approaches

Before describing the approach proposed here some additional alternatives are mentioned. There are, of course, numerous approaches for distributed systems in general and distributed object systems in particular. Their discussion is beyond the scope of this thesis[13]. An approach considered during the evaluation of various alternatives was an implementation upon HTTP[14] as discussed in (Včkovski, 1995). HTTP is used to transmit method invocations to the server and return the results to the client. An HTTP-server (i.e., WWW-server) plays the role of the object implementation. Methods invoked on objects are encoded as URL as

> http://server/object-reference?method+arg1+..+argn
>
> Example:
> http://foo.bar.com/climate/temp?value+long=45.12+lat=41.2

Using the typing-mechanism provided by MIME (Borenstein & Freed, 1993), the methods could return the results as any of the predefined types. This means that there is a kind of automatic marshaling.

Such an approach has the advantage of a simple implementation on the server side. There are many HTTP-server programs available, and most of them can be easily extended using the so-called *Common Gateway Interface (CGI)*. This allows programs external to the server to handle certain requests to the server, i.e., URLs. In the example above, the server could have been configured to call a program `temperature` within the `climate` directory. The additional information within the URL (i.e., method name and arguments) is passed to the program called in a way specified by CGI. Based on this information, the `temperature` program can construct the request's result in any format provided that there is a corresponding MIME type. Since the server-called program is also informed of the *type* of client issuing the request it can flexibly return various results of the request, depending on the client. For example, if the request were issued from a VDS-aware application the program would return the result encoded in an agreed format that would be used in that data user's application. If the request were issued from a standard HTML-viewer, i.e., WWW-browser, the result returned could be in the form of a simple text representation or as an image displayable in the browser, e.g., a GIF-image (see Figure 7.4).

The implementation of VDS-aware client applications, however, is not so simple because the client application has to handle HTTP communication[15]. This means that there is a certain programming overhead involved when using a VDS. This overhead could be reduced by providing a small (class) library of

---

[13] A good overview is given in (Coulouris *et al.*, 1994). See also the description of a recent environment called *Inferno* (Lucent, 1996).

[14] Hypertext Transfer Protocol, see page 134.

[15] On the server side, this is handled by the HTTP server directly without intervention of the programs called through CGI.

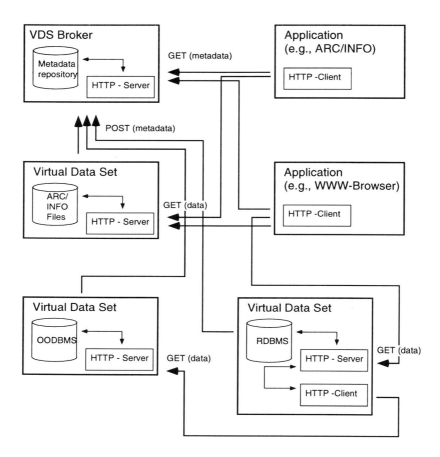

Figure 7.4: VDS implementation using HTTP

functions (objects) that can be used to encapsulate the communication details of a method invocation. As HTTP operates on TCP streams (i.e., sockets as discussed on page 129) such a library would be relatively simple.

Within this thesis, however, a different approach is proposed. It is based on *Java*, both as a programming environment for VDS and as a software system for distributed objects. The next section will therefore describe these aspects of Java in more detail.

7.5 Java

7.5.1 Overview

Java is both an object-oriented programming language *and* an environment for distributed computing. Java had its origin in 1990 in a research project headed by James Gosling at Sun Microsystems. The project aimed at developing new techniques for programming in the domain of consumer electronics, e.g., video recorders, washing machines, television sets and so on. Software for consumer devices, sometimes called *embedded systems*, have some specific requirements (Flanagan, 1996, p. 4):

- Software for consumer electronics has high portability requirements. The microprocessors used in these systems are changed often because new chips become more cost-efficient or introduce new features. The software needs therefore to be ported often to new platforms.

- There are also high reliability requirements. If the software fails, the manufacturer has typically to replace the entire device. Consumers do not expect software malfunction on such devices and there are typically also no recovery mechanisms other than a power-off-power-on cycle.

- Embedded systems often require a multi-threaded[16] environment. For example, a video recorder playing a video tape needs to perform various partially independent tasks in parallel, such as controlling the motor, monitoring the buttons for user interaction, displaying status information and so on.

Embedded systems are traditionally programmed using a processor-specific assembler or – if available – C-compilers[17]. To overcome the limitations of using processor-specific environments a new programming language (eventually called *Java*) was designed that would be more appropriate for the programming of embedded systems in consumer electronics. The language was architecture-neutral and based on concepts to enhance reliability and allow simple multi-threading.

---

[16]Multiple lines of independent dynamic action within an application (Firesmith & Eykholt, 1995, p. 443).
[17]In most cases *cross-compilers* are used which allow development on personal computers or workstations and produce processor-specific code.

As the World-Wide Web started to become one of the major applications on the Internet, it was soon realized that Java would be an ideal platform for programming within the Internet, particularly because of its ability to run on many platforms. The team on the Java project therefore developed a Web browser called *HotJava*. This application had several interesting features (Gosling & McGilton, 1995, pp. 76–79):

Applets HotJava introduced the notion of applets. These are *mini-applications* that run within other applications. With HotJava a few extensions to HTML were introduced which allowed one to specify an applet to be responsible for the visual rendering of a rectangular area within an HTML document. Since an applet is "real code", this allows arbitrary kinds of user interaction and dynamics. This feature of HotJava – quickly picked up by mainstream Web browsers – was certainly one of the most important reasons for Java's success[18].

Dynamic types HotJava had a dynamic extension mechanism which allowed the browser to be extended with new functionality *at runtime*. This was used to implement dynamic MIME type handling[19]. If HotJava encounters an unknown MIME type as a result of a request, it tries to dynamically load a handler object for that specific MIME type. A content provider with some specific type of information, e.g., digital maps, could provide the necessary handler objects which would allow users to access the data. It is of key importance that such objects are loaded upon request and bound *dynamically* into the browser.

Dynamic protocols The dynamic extension mechanism is also used to extend the browser at runtime by new *protocol* handler objects. As was shown on page 135, the World-Wide Web uses URLs for resource identification. These URLs contain amongst other things a protocol identification. HotJava can be extended at runtime by new protocol handlers as soon as an unknown protocol is encountered.

To sum up, HotJava had as a core technology a dynamic extension mechanism which allowed small parts of code to be transferred over the network and bound to the application at runtime. This technology is made possible by various features of the Java software system as will become apparent in the following sections.

Since its release to the public in early 1995, Java has become very popular, mostly because of its ability to spice up Web pages with dynamic content such as animations. The further development of Java benefited from its popularity within the Internet. However, the popularity did also freeze the development in some areas since Java was used by many people, and numerous vendors

---

[18] As a consequence of this, Java is still often misleadingly only identified as a language to create applets ignoring its potential, which covers many other application areas.

[19] See also discussion on content negotiation on page 135.

have licensed the technology from Sun Microsystems. This is apparent in some heated debates going on about possible extensions and missing features. Such a development has been seen in the early stages of all programming languages gaining a certain interest when programmers switching from a previous programming environment miss their favorite feature within the new environment.

Java is more than a *programming language*. Java also defines a *common class library* and a *runtime environment*. This is somewhat similar to the Oberon system (Wirth & Gutknecht, 1992) although Java's runtime environment is not on the operating system level as it is in Oberon. In the following, the design goals of Java are discussed first, followed by the description of the main features and Java as a distributed computing environment.

7.5.2 Design goals

The Java programming language was designed along requirements derived from the consumer electronics environment. It was clear that traditional programming languages such as C or C++ posed many problems regarding portability issues, such as the key role of *pointers* in C/C++. It was soon realized that these problems could be best addressed by creating an entirely new language environment (Gosling & McGilton, 1995, p. 12). The design and architecture of the new language Java were able to benefit from many programming languages that had been around for a while, such as the Cedar/Mesa-system (Teitelmann, 1984) or UCSD-p Pascal (Wirth, 1996). The design goals of Java will be presented briefly, following the discussion in (Gosling & McGilton, 1995, p. 13–15):

Simplicity The Java language is *simple* so that it can be used without extensive programmer training.

Object-orientation Java fully adopts object technology. It has been seen in the last few decades that object-oriented design and programming is an efficient tool to manage or reduce the complexity of software engineering. Moreover, the message-passing paradigm[20] fits the the requirements of distributed client/server systems very well.

Familiarity Java has adopted many of the syntactical and semantical elements of C++. Programmers with C++ experience will therefore experience steep learning curves.

Robustness The high reliability requirements need concepts to support programmers in creating reliable software. Java's features enhancing robustness are a *declarative exception handling* mechanism, *automatic*

---
[20] In the context of object-oriented systems, "message" is synonymous to "method call", "request" or "operation invocation" (Firesmith & Eykholt, 1995, p. 245). A message sent to an object means the invocation of the corresponding operation in CORBA terminology, i.e., call of the corresponding method.

memory management which make pointers obsolete, and an extensive compile-time and runtime type checking.

Security Security is always an issue in distributed systems. Software architectures allowing the local execution of code that is dynamically loaded into the local system from a remote site need to be very carefully designed to avoid invasion by malicious programs. Java includes several security levels to prevent tampering of code by viruses, to forbid access to local system resources such as the file system and so on.

Architecture neutrality Being platform-independent was one of the major objectives for Java. Java achieves this independence by introducing an intermediate binary representation of compiled Java programs. Programs are compiled into a *byte-code* which is then interpreted by a platform-specific interpreter.

Portability Portability means not only architecture neutrality. Java also exactly defines all the basic data types and the behavior of the operations between them. The sum of any two numbers is the same on every Java system, something that is not at all common in other environments due to differences in the number's binary representation.

Multi-threading Java is designed for multi-threading from ground up. This means that there is a simple semaphore-model[21] built into the language allowing object methods to be synchronized. The class library offers primitives for creating own threads and thread groups. All system libraries in Java are *thread-safe* and re-entrant[22].

Dynamic linking Dynamic extension of an application at runtime requires dynamic linking to be available in order that new modules can be bound to the application. The dynamic linker resolves references only when needed, i.e., classes never needed in an application are not necessarily loaded into memory at startup-time.

Reusability The experience with object-oriented systems has shown that major problems when reusing existing code are caused by clashing *names*, despite all the encapsulation and so on that was achieved by virtue of object orientation. This means that there might be the same names identifying, e.g., classes, in different class libraries. A class library designer needs to be free to name his or her elements but also to be guaranteed that the names will not conflict with those in other class libraries. Java uses so-called *packages* to provide unique name spaces.

---

[21] Actually, a binary semaphore model ("mutex"-semaphore).

[22] This means that a function can be called by another thread while it is executing. Such functions need to be careful about static variables that might be modified by various simultaneously running threads of the function.

Data type	Length in bits	Domain
byte	8	integer[a] $[-2^7, 2^7 - 1]$
short	16	integer $[-2^{15}, 2^{15} - 1]$
int	32	integer $[-2^{31}, 2^{31} - 1]$
long	64	integer $[-2^{63}, 2^{63} - 1]$
char	16	Unicode[b] character
float	32	IEEE 754 real number
double	64	IEEE 754 real number
boolean	n/a	true or false[c]

[a]All integer types are *signed*.
[b]Using 16-bit Unicode characters is important for internationalization and a fundamental difference from C/C++.
[c]In Java, boolean types cannot be converted directly to any numeric type.

Table 7.1: Primitive data types in Java

7.5.3 Main features

The discussion of all features of Java is beyond the scope of this overview[23]. Here, only some of the features important for the understanding of the design presented in chapter 8 on page 181 are described.

Primitive data types

Unlike other fully object-oriented systems, Java defines some data types which are *not* objects. This breaks the rule for a pure object-oriented language but has been introduced for performance reasons. However, there are object-wrappers for all primitive data types available within the class library. The primitive data types are shown in table 7.1.

Arrays are not primitive data types in Java. They are contained as first class objects in the class hierarchy. This enables both the compiler and the runtime system to perform tasks such as checking for out-of-bounds indices, dynamic allocation of array elements and more. However, Java defines enough special syntax for arrays that it is still useful to consider them a reference type different from object references (Flanagan, 1996, p. 32).

---
[23]For a comprehensive discussion see (Flanagan, 1996), (Gosling & McGilton, 1995) or (van Hoff *et al.*, 1996).

Inheritance and class hierarchy

Java's object model uses a single inheritance model, i.e., a child is constructed in terms of the definition of only a *single* parent (Firesmith & Eykholt, 1995, p. 209). Moreover, every class in Java has a common ancestor, the base class Object. This means that there is one single acyclic inheritance graph, i.e., an inheritance tree, as opposed to the case in C++ where every class can live in a multiple inheritance relationship with various roots, i.e., in an "inheritance forest". Having only single inheritance avoids the problems found with multiple inheritance such as a common ancestor in two superclasses of a class. Consider for example class D being derived both from class B and class C. Both B and C shall be a specialization of the base class A. Let's assume that both B and C override a method x defined in A (x_B and x_C). Which method implementation should now be called when invoking x on D? In C++, the only way to avoid this problem is either to explicitly specify *which* x to use or to define the base class A to be virtual when deriving B and C. This in turn disables overwriting of A's methods in B and C and makes sure that there is only one copy of A in D.

While single inheritance avoids these problems, there are often situations where multiple inheritance might be desired, i.e., when a class needs to be a specialization of two independent base classes. The only way to achieve this in Java is through *interfaces* as will be discussed below. Having Object as a common base class for *every class* simplifies the implementation of many helper classes such as *containers*. A container capable of holding references to Object can hold *any* object type. The trade-off is, however, that there is no sensible type-checking possible any more since any object can be assigned to, for instance, a list element even if the list is thought to hold only objects of a specific type. This is a major criticism in Java's design (Fischbach, 1996) and there are already attempts to overcome this limitation by introducing *parameterized types* similar to *templates* in C++.

The base class of the hierarchy Object defines a set of useful methods than can be used within all classes. Its definition is shown in listing 7.2. Note that there is a method clone() defined which is used to copy instances. There seems to be a semantical inconsistency in Java in that assignment of a value of a primitive data type to a (compatible) type copies the value, where as assignment of objects does not copy the objects. However, this is actually no inconsistency because Java only manages object *references* (see listing 7.3).

Interfaces

Java's lack of multiple inheritance is compensated by the concept of interfaces. Interfaces are a kind of *abstract data type* similar to *protocols* used in Objective C (Pinson & Wiener, 1991) and the interfaces seen previously in the discussion of CORBA's IDL on page 150. Interfaces define a set of methods but do not carry any implementation. A class can *implement* a set of interfaces. If a class implements an interface, this means that it con-

Listing 7.2 Java base class `Object`

```
public class Object {

  // Returns the Class of this Object
  public final native Class getClass();

  // Returns a hashcode for this Object.
  public native int hashCode();

  // Compares two Objects for equality.
  public boolean equals(Object obj);                                    10

  // Creates a clone of the Object.
  protected native Object clone()
    throws CloneNotSupportedException;

  // Returns a String that represents the value of this Object.
  public String toString();

  // Notifies a single waiting thread from another thread.
  public final native void notify();                                    20

  // Notifies all waiting threads for a condition to change.
  public final native void notifyAll();

  // Causes a thread to wait until notification or a timeout.
  public final native void wait(long timeout)
    throws InterruptedException;

  // More accurate wait.
  public final void wait(long timeout, int nanos)                       30
    throws InterruptedException;

  // Causes a thread to wait for ever until it is notified.
  public final void wait() throws InterruptedException;

  // Code to perform when this object is garbage collected.
  protected void finalize() throws Throwable;
}
```

Listing 7.3 Assignment of primitive data types and object references

```
int a = 2;
int b;

// a copied to b
b = a;

// value of a does not change
b = 4;

MyClass foo, bar;                                                       10

// create new instance of MyClass
// and assign reference to 'foo'
foo = new MyClass();

// bar references same instance as foo
// (object reference is copied)
bar = foo;

// bar is now a copy of foo (i.e., new instance)                        20
bar = foo.clone();
```

tains all the methods specified in that interface. Moreover, interfaces can also be inherited from other interfaces, i.e., there is an interface inheritance tree. Consider for example the program shown in listing 7.4. It declares three interfaces, **interfaceA**, **interfaceB** and **interfaceC**, where **interfaceB** is inherited from **interfaceA**. Class **someClass** directly inherits from **Object**[24] and implements all three interfaces. In the static method **main()** of class **interfaceExample** an instance of **someClass** is created and assigned to *interface references* **a_ref** and **b_ref**. An interface reference is a reference to an object which implements that interface. Consequently, it is possible to call the methods declared in the interface through the interface reference, but not any other method of the object.

This interface scheme is extremely powerful especially in the context of interoperability. Interfaces can characterize particular "abilities" of objects. For instance, one could define interfaces that encapsulate methods for *printing, serializing, displaying*. If any class implements the **printable** interface, the containing application knows that it can call a set of methods of that object which are used to render a visual representation of the object on a printer. An application interested only in printing does not need to be aware of the other interfaces also implemented by the class. The drawback of using interfaces, i.e., declarations *without* implementation, is that in the example above

---

[24] If a class does not specify a superclass then **Object** is implied as direct superclass.

Listing 7.4 Interfaces in Java

```java
interface interfaceA {
  int foo();
}
interface interfaceB extends interfaceA {
  String bar();
}
interface interfaceC {
  double foobar();
}
class someClass implements interfaceB, interfaceC {
  // method declared in interfaceA
  public int foo() {
    return 42;
  }
  // method declared in interfaceA and inherited in interfaceB
  public String bar() {
    return "bar!";
  }
  // method declared in interfaceC
  public double foobar() {
    return 3.14159;
  }
}
public class interfaceExample {
  public static void main(String args[]) {
    // create instance
    someClass aclass = new someClass();

    interfaceA a_ref = aclass;  // references to interfaces
    interfaceB b_ref = aclass;

    int a = a_ref.foo();        // foo through interfaceA
    int b = b_ref.foo();        // foo through interfaceB
    String c = b_ref.bar();     // bar through interfaceB
                                // a_ref.bar() is illegal!
    int d = aclass.foo();       // foo from someClass
    String e = aclass.bar();    // bar from someClass
    double f = aclass.foobar(); // foobar from someClass
  }
}
```

all classes implementing an interface must actually provide implementations of all methods. However, this is sometimes a nice side-effect in that users of certain interfaces can be *forced* to provide a complete implementation. The declarative exception handling mentioned before also applies to interfaces. Every method in an interface can be declared to throw (i.e., stop execution and notifiy the caller) one or more exceptions. Interface users *must* either catch the exception or declare themselves to (possibly) throw the exception. This type of exception handling possibly enhances software reliability since it *forces* object or interface users to handle exceptions[25].

Memory management

The availability of a powerful memory management system is of paramount importance for reliable implementations. In traditional software environments most of the fatal runtime errors are somehow related to memory management, such as:

- Languages such as C/C++ provide *pointers* for indirect[26] access. It is *the programmer's responsibility* to make sure that the value stored in a pointer variable references a valid memory location (i.e., no *dangling* or *null pointers*). Systems with memory protection will stop the execution of programs accessing illegal memory locations while systems without protection will eventually hang.

- *Dynamic memory allocation* is a key element of good programming style. Benign applications allocate as much memory as they need in order to make best use of system resources. Dynamic memory allocation, however, also leads to many possible programming mistakes, such as for example *memory leaks*. This happens if a previously allocated portion of memory is not used any more but not disposed of correctly. For applications with short execution times this is usually no problem because the operating system typically re-collects any memory allocated by an application. However, applications that are planned to run for long times – such as for example a server within a client/server environment – may exhaust the system's memory resources by memory leaks.

- *Multi-threading* typically makes memory management more difficult. Certain memory locations might be shared between various threads.

---

[25] It is possible, of course, to propagate exceptions automatically to the next higher level, i.e., next stack frame, without any processing. Since the exceptions are *language* elements, however, the compiler can already check *whether* exceptions are caught or not and point the programmer to possible omissions.

[26] Pointers are often misleadingly thought to be *direct* access to memory because they allow direct manipulation of memory locations. However, the access is – strictly defined – indirect. Access to memory happens via one indirection step by virtue of a memory address stored in the pointer variable, whereas normal variables allow direct access to the memory because they are bound to a specific memory location by the compiler or the program loader's relocation mechanism.

It is therefore not trivial to determine the overall lifetime of an object in memory. A single thread cannot simply dispose a memory location that is possibly also used by another thread.

- Memory addresses are a highly platform-specific entity. Code that directly manipulates addresses (e.g., using C/C++ pointer arithmetic) is likely not to be very portable across various platforms.

- Direct access to memory can also impair the system's security[27].

Java incorporates its own memory management which allows dynamic allocation but avoids direct memory references, i.e., there are *no pointers* in Java. Objects are accessed through so-called *references*. References cannot be manipulated using any arithmetic operation. A reference can be either assigned to an *existing* instance of an object or to null, which is a unique unresolved reference. The only operator for references besides assignment is the comparison for equality (== operator) which checks if both references point to the same object.

Dynamic memory management allows the application to create new object instances dynamically at runtime but not to delete them. This is taken over by an automatic *garbage collector*. It is never necessary to reclaim memory not used any more by the programmer. Instead, the garbage collector running as a low-priority thread frees memory occupied by objects that are not referenced any more, i.e., there is no reference to the object in any running thread. The Java runtime system is able to accomplish this since there is only one way to create a new reference, namely by assignment. Therefore, it is possible to track the existing references to every object.

This scheme offers also useful future extensions for, e.g., *caching* within applications. Consider for example an environment with potentially high memory requirements such as in many applications within spatial data handling. During a session many queries and intermediate results might be created. For many of these intermediate results it is known how they can be re-created. In other words, it is feasible to free the memory occupied if necessary because the result can be re-created on demand, i.e., keeping those objects in primary memory is only a matter of efficiency. An extended garbage collector could, for instance, delete objects tagged as "re-creatable" if more memory is required *without* removing the references to the object. As soon as it is accessed again the object could be re-created transparently by the runtime environment. As many data as possible would therefore be kept in primary memory without the user having to discard intermediate data explicitly.

---

[27]One might argue that this is, if at all, a problem of the operating system. Security considerations are, however, important in the context of distributed systems as will be seen later on, and it is generally a good security policy to provide as much security as possible at each level, i.e., not to omit possible security actions only because the actions should actually be taken by someone else.

Class libraries

As was mentioned before, Java allows one to define *packages* with private name spaces. There is a set of basic packages which are part of the Java environment. Parts of this class library are tightly bound into the Java environment. For example, key elements such as the class hierarchy root `Object` and other fundamental components are contained in the package `java.lang`. On the one hand it is somewhat confusing to have imperative language components within a class library package, since class libraries are often understood as a useful but optional entity. On the other hand it is often useful to have access to the core language elements by the same mechanisms as other elements are accessed, i.e., there is no need for special syntactical constructs. However, Java treats the package `java.lang` in a special way in that it is always implicitly imported[28] in any program. Table 7.2 shows the packages contained in the basic Java environment.

Main differences from C/C++

The main differences from C++, which is currently the most widely used object-oriented language in the programming mainstream, are a kind of a dialectical summary of the features of the Java language (see below). Objectively comparing programming languages is usually a very difficult if not impossible task. In most cases, it is not only the programming language itself that needs to be taken into consideration, but also its context, i.e., tools for the programmer such as compilers and debuggers, libraries, runtime environments like Java's (see section 7.5.4) and so on. The main differences presented here are not at all exhaustive and are intended only to contrast some techniques routinely used in C++ development with Java. The list below as inspired by Gosling and McGilton (1995, p. 26) shows merely the features *removed* from C/C++:

- Java does not rely on a preprocessor. In C/C++ a preprocessor is typically used for defining constants, including secondary source files ("include files", e.g., with variable and function declarations) and conditional compilation. Making no distinction between *declaration* and *definition* in Java makes include files obsolete. Furthermore, Java allows constants to be defined as constant variables. Conditional compilation in Java is dependent on the compiler's ability to detect "dead", i.e., never reached code, which is bracketed out by conditional statements with constant logical expression.

- Java has no "stand-alone" functions. Every procedural component needs to be embedded in an object. It is a philosophical question whether it

---

[28] *Importing* means in Java that all classes and interfaces defined in the package are directly visible without explicit specification of the package name. For example, class `String` which is defined in package `java.lang` can be directly referenced by `String` instead of the fully qualified name `java.lang.String`.

Package	Description (Gosling & McGilton, 1995, p. 16)
java.lang	Main elements of the Java language, such as Object, Class, String, threads, wrappers for primitive data types and so on.
java.io	Streams and random-access files.
java.net	Berkeley sockets and URLs.
java.util	Containers and other utility classes.
java.awt	The *Abstract Windowing Toolkit* (AWT) which provides platform-independent access to the system's windowing system in order to create graphical user interfaces.
java.awt.peer	Peer classes of AWT making the connection to the system's native windowing system.
java.awt.image	Utility classes providing simple image processing capabilities to the AWT.
java.applet	Classes for implementing *applets* and their container.

Table 7.2: Java packages

is sensible to have operations independent of any object instances or classes or if every operation must be associated with an object or class. Java does support class-level methods, i.e., methods that are not bound to a specific object instance.

- Java does not support multiple class inheritance as discussed before.

- There is no `goto` statement in Java.

- Java does not provide *operator overloading*. This is a syntactical but not a semantical difference. Operator overloading in C++ allows a special "meaning" to be assigned to the language's standard operators such as +, -, == and so on. A C++-compiler translates an expression a+b into a method call `a.operator+(b)`, i.e., operator overloading does not provide additional functionality but it provides more elegance and readability.

- There are no *pointers* in Java.

- There is no *automatic coercion* from data types of higher precision into data types of lower precision as happens in C/C++, for instance, when assigning a `double` value to an `int` value.

7.5.4 Java as a distributed computing environment

Overview

Java is more than just a programming language. Based on the requirements for portability there are some characteristics making Java very suitable for the implementation of distributed systems (e.g., the Caltech Infospheres Project (Chandy & Rifkin, 1996)), such as:

- Java is not compiled directly into machine code but into an intermediary *byte code* which is then interpreted on the target platform. Java code can therefore be easily migrated from platform to platform.

- Java offers powerful *dynamic linking* (late binding), making it possible, for example, to load code at runtime and bind it to the application.

- *Interprocess communication* as provided by the Berkeley sockets (see section 6.3.2 on page 129) is available on a high abstraction level (i.e., easy-to-use) within the package `java.net`.

- *Security issues* are of paramount importance in distributed environments.

Java virtual machine

Java defines a *virtual machine* with a series of operation codes. A Java *compiler* translates a Java program into instructions for the virtual machine (VM). A Java *interpreter* (or code generator) translates these instructions in turn into platform-specific machine code. This is the same approach as was used, for example, in the family of UCSD p-system languages (Wirth, 1996) and somehow incorporates the benefits of both a compiled and an interpreted environment[29]. By compiling a program into VM instructions, it is possible to keep the interpreter simple. It has been shown in the POSTSCRIPT case study that one of the major drawbacks of using POSTSCRIPT programs as a flexible data carrier is the complexity of the POSTSCRIPT interpreter. An interpreter for VM code needs only to understand a "handful" of instructions. The *lexical scanning* and *parsing* steps are taken over by the compiler. Moreover, the system generics are based on the POSIX standard for operating system calls which is supported by many current platforms, making the port of the VM interpreter to a new platform straightforward.

A typical cycle is shown in figure 7.5. The Java compiler creates a binary representation of all the classes[30] within a program in VM byte code. Execution of the programs loads the classes into memory and passes them first into a *byte code verifier* which makes sure that the classes do not perform illegal tasks (see security discussion below). After verification the byte code is turned into executable machine code and passed to the processor. There are two important steps to note here:

- The loading of the byte code can happen from *any data source* and is dependent on the byte code loader. The data source could be a file on the local file system, a remote network resource, a database management system, and so on. A Java application can provide its own byte code loader as a specialization of class `ClassLoader` in the package `java.lang`[31].

- The method of code generation is not pre-determined. A simple code generator is a step-by-step interpreter of the VM instructions, while a more sophisticated code generator would compile the entire unit into machine code after verification and then pass it to the hardware, boosting the performance[32].

Listing 7.6 shows the mnemonics of the byte code generated for the method `main()` shown in listing 7.5.

---

[29] It is important to make the difference between *language* and *environment*. A programming language *per se* is not interpreted, but its implementation might work as an interpreter. However, certain language designs are more easily implemented as interpreter than as compiler.

[30] There are *only* classes in Java, not stand-alone functions.

[31] See also example E34 discussed in section 8.3.3.

[32] This is sometimes called a *just-in-time* compiler.

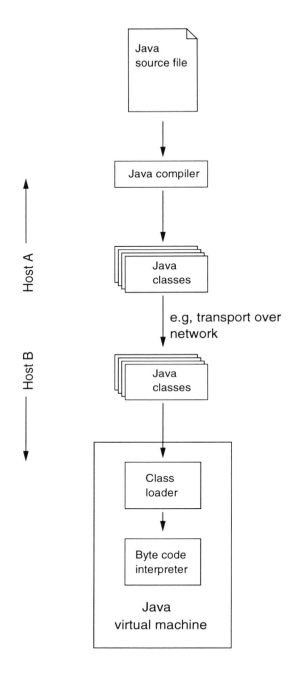

Figure 7.5: Java program compilation and execution

Listing 7.5 Method main() of byte code in 7.6

public class simple {

  public static void main(String args[]) {
    int a=1, b=2, c;
    c = a+b;
    System.out.println("Result is "+c);
  }

}

Listing 7.6 Byte code generated from 7.5

```
 0 iconst_1
 1 istore_1
 2 iconst_2
 3 istore_2
 4 iload_1
 5 iload_2
 6 iadd
 7 istore_3
 8 getstatic #10
    <Field java.lang.System.out Ljava/io/PrintStream;>
11 new #5
   <Class java.lang.StringBuffer>
14 dup
15 invokenonvirtual #13
   <Method java.lang.StringBuffer.<init>()V>
18 ldc #1
   <String "Result is ">
20 invokevirtual #12
   <Method java.lang.StringBuffer.append(Ljava/lang/String;)
    Ljava/lang/StringBuffer;>
23 iload_3
24 invokevirtual #7
   <Method java.lang.StringBuffer.append(I)
    Ljava/lang/StringBuffer;>
27 invokevirtual #11
   <Method java.lang.StringBuffer.toString()
    Ljava/lang/String;>
30 invokevirtual #9
   <Method java.io.PrintStream.println(Ljava/lang/String;)V>
33 return
```

Dynamic linking

References to other classes and their methods are contained as string symbols in a compiled Java class as can be seen in the example discussed above (see listing7.6). These references are resolved at runtime by the byte code loader. The byte code contains also the definition of the class, i.e., instance and class variables, the signatures of all methods, the class's superclass and all interfaces implemented. This mechanism has many advantages:

- Applications can find out which methods a *compiled* class contains and which interfaces it supports, somewhat similar to the POSTSCRIPT Document Structuring Conventions discussed in section 6.2 on page 119.

- References can be resolved only when needed. The runtime system might for example resolve classes with no static members only when the first instance is created.

- An application can be extended with new classes at runtime.

The dynamic linking using symbolic references also solves a problem known as the *fragile superclass problem* in C++ (Gosling & McGilton, 1995, p. 53). This arises when a superclass is modified in a way that changes the memory layout of the class, e.g., by adding new instance variables or virtual methods. In such a case, every derived class of this superclass as well as any other code that references this class needs to be recompiled since the references depend on the memory layout of a compiled class. Within large projects involving many developers, this might lead to situations where base classes are artificially "frozen" in order to reduce side-effects.

The dynamic linking is shown in a simple example in listings 7.7 to 7.9. It consists of two classes, SomeClass and ClassUser and a common interface SomeInterface. The method main() in class ClassUser *dynamically* loads the class SomeClass and calls a method through the interface SomeInterface. It is important to note that class ClassUser does not need to have any other additional information about class SomeClass other than that it implements interface SomeInterface. This is checked dynamically in the example as well. In other words, the class SomeClass could be a VDS which is accessed through a well-defined interface SomeInterface. The data user does not need to know any intrinsics of the class used. It is also not necessary to have any source code representation of it or of the interface, as is, for example, the case with COM.

Interprocess communication

Interprocess communication is a core component of every distributed system. Java – sometimes called a "network programming language" – contains a simple socket interface in its base class library package java.net. In combination with other elements of the class library such as the streams provided in

Listing 7.7 Common interface `SomeInterface`

```
public interface SomeInterface {

  // print the string to somewhere
  public void printString(String astring);

}
```

Listing 7.8 `SomeClass`, implements `SomeInterface`

```
public class SomeClass implements SomeInterface {

  // instance vars
  int dummy = 42;

  // print the string to standard out
  public void printString(String astring) {
    System.out.println(astring);
  }

}
```

Listing 7.9 Using `SomeClass` through interface `SomeInterface`

```java
public class ClassUser {

    // we do not want to deal with exceptions within the method,
    // therefore we have to declare that we throw any exception
    public static void main(String args[]) throws Exception {

        // dynamically load a class by name
        Class someclass = Class.forName("SomeClass");

        // create new instance
        Object o = someclass.newInstance();

        if (o instanceof SomeInterface) {
            // we know that someclass implements SomeInterface,
            // so it is safe to cast here
            SomeInterface interf = (SomeInterface)o;

            // call method printString in object o
            interf.printString("Hello, World!");
        }
    }
}
```

`java.io` and Java's multi-threading facilities, this offers a flexible and easy-to-use way of network programming. Multi-threading is thereby a key issue in a classical client/server model. The server has to listen for incoming connections. If a connection is established, the server should both continue to serve this connection and to listen for new connections simultaneously. This is particularly important if the service provided by the server typically leads to long connections such as, for example, a `telnet`-session. In a classical single-threaded environment the server process would, as soon as a request is coming in, spawn a child process and let the child process manage the communication with the client while the parent process continued to listen for incoming requests, i.e., there is one child for every client connection.

In Java, this is implemented in a similar fashion, only no other processes need to be spawned. Client connections can be managed by separate threads within the same process. Listing 7.10 shows a simple implementation of a server which does nothing but echo whatever the client sends back to the client (derived from (Flanagan, 1996, p. 145)). The brevity of the code demonstrates the power of the networking class libraries. Client programming is usually less complex concerning the communication because there is no need for multiple threads. Listing 7.11 shows the client counterpart. The client reads lines from standard input, e.g., the terminal, and sends whatever the user types to the server and displays everything received from the server on the terminal. This is basically the functionality of a `telnet`-client.

At the time of writing, there are also methods for interprocess communication being developed on an abstraction level higher than Berkeley sockets, such as a CORBA binding and a somewhat simpler mechanism for *remote method invocation*.

Security

Security has always been a major concern in distributed systems and may be the topic most discussed about Java. In fact, the ease of code migration increases problems such as computer viruses. There are several security areas possibly affected by "moving code":

Destruction The code could destroy local information such as deleting local files etc.

Espionage The code could retrieve information from the local system and send it to another remote node.

Denial of service attacks The code could maliciously consume system resources such as memory and force the local system to its knees.

Misinformation The code could maliciously produce wrong results. For example, if the code is a type of calculator it could introduce random errors when performing operations that are not easy to comprehend manually.

There are two strategies in Java to prevent such attacks. The first one deals with destruction and espionage and is performed after loading the byte code. The byte code verifier (see figure 7.5) uses an instance of Java's `SecurityManager` class to perform security checks on the byte code. The security manager implements a security policy by defining what the code is allowed to do. A security manager within a Web browser, for example, is typically far more restrictive than a stand-alone, unnetworked application. *Applets* within Web-pages are usually restricted in a way that they cannot access local resources (read files and directories etc.), cannot load classes themselves, execute other processes, and so on.

The second security strategy applies to the second group of malicious attacks. Such attacks cannot be detected automatically since the computer cannot determine if a result is correct or if the resource allocation is in the interest of the user or not. The strategy pursued here is based on a *certification* of code from remote sites. A certification consists of a digital signature and corresponding cryptological checksum. This allows the local user to determine whether the code has been modified during travel time and whether the sender is trustworthy. In traditional environments, we trust the software's "correctness" because we know who produced it and, based on the software's packing, that we really got the software from the manufacturer. In

Listing 7.10 Socket server in Java

```java
import java.io.*;
import java.net.*;
public class SimpleServer extends Thread {
  final int port = 8000;
  public static void main(String args[]) {
    new SimpleServer();
  }
  public SimpleServer() {
    start();
  }
  public void run() {
    ServerSocket listenTo;
    try {
      listenTo = new ServerSocket(port);
      while (true) {
        Socket clientsocket = listenTo.accept();
        new SimpleConnection(clientsocket);
      }
    } catch (IOException e) {};
  }
}
class SimpleConnection extends Thread {
  Socket client;
  public SimpleConnection(Socket aclient) {
    client = aclient;
    start();
  }
  public void run() {
    try {
      DataInputStream din =
        new DataInputStream(client.getInputStream());
      PrintStream dout =
        new PrintStream(client.getOutputStream());
      while ( true ) {
        String s = din.readLine();
        dout.println("Echo:  "+s);
      }
    }
    catch (IOException e) {}
    finally {
      try { client.close(); } catch (IOException e2) {}
    }
  }
}
```

Listing 7.11 Socket client in Java

```java
import java.io.*;
import java.net.*;

public class SimpleClient {
  final static int port = 8000;

  public static void main(String args[]) {

    String host = args[0];
    Socket socket = null;                                          10
    try {
      socket = new Socket(host,port);
      DataInputStream din =
          new DataInputStream(socket.getInputStream());
      PrintStream dout =
          new PrintStream(socket.getOutputStream());
      DataInputStream stdin =
          new DataInputStream(System.in);

      while (true) {                                               20
        String line = stdin.readLine();
        if (line == null) break;
        dout.println(line);
        line = din.readLine();
        if (line == null) break;
        System.out.println("From server:   "+line);
      }
    }
    catch (IOException e) {}
    finally {                                                      30
      try { socket.close(); } catch (IOException e2) {}
    }
  }
}
```

distributed, component-based systems it is very likely that software will be used of vendors that are not known. Moreover, the trustworthiness cannot be estimated on the basis of external factors such as quality of the packing, quality of the documentation and distribution media (e.g., "is there a nice label on the diskette?").

In the context of VDS, it is also very important to provide corresponding security mechanisms. Data users of traditional, static data formats did not have to worry about viruses in their data sets (although they had to worry about wrong data). Providing certification allows data users to trust a traditional data set or a VDS as much as they trust the data producer.

7.6 Review

All features of the Java system discussed in the previous chapter help the implementation of VDS with Java. As a summary, the feasibility of Java is based on:

- Java is available on many different platforms with identical execution semantics (for instance, data types are the same on all platforms).

- Java offers various techniques for the implementation of distributed systems, such as interprocess communication, CORBA binding, remote method invocation (RMI), dynamic loading of classes from any data source (e.g., network connection).

- Java is a pure object-oriented environment which is simpler than C++ (fewer features) but more robust due to features such as automatic garbage collection and strong type checking.

- Java uses the notion of *interfaces* for implementation-free classes. The interfaces offer a natural implementation alternative for the *well-known* interfaces required by VDS and OGIS.

- Classes can be introspected[33] to assess information about a class's structure, which can be thought of metadata in the context of a VDS.

- Java offers security schemata which are a must in a distributed system.

The following chapter demonstrates these features with some exemplary implementations. The examples are deliberately kept simple in order not to obscure the essential details.

---

[33]The Java beans specification (Javasoft, 1996) substantially enhances this capability.

CHAPTER EIGHT

Examples

8.1 Introduction

This chapter uses the findings of chapters 6 and 7 for a few examples regarding the implementation of the essential models of fields (chapter 4) and uncertain values (chapter 5). The objective of these examples is to show some of the aspects of an implementation in Java, the advantages and shortcomings.

Interoperability based on interfaces cannot be implemented by a *single* instance. There need to be at least two interoperating components, typically one that *implements* an interface and another one that *uses* an object through that interface. The range of possible applications "on both sides" of the interface is unlimited. In a traditional geoprocessing application, the object implementing an interface would represent a dataset (i.e., a VDS) and the application accessing that object could be a graphical viewer. A more advanced application could consist of an object representing, for instance, a mathematical simulation model on the one side (which itself uses again other objects, see also figure 3.4 on page 59). On the other side, the data consumer application could be a simple converter object, that writes a VDS's content to a disk file in a standard format.

The examples in this chapter concentrate mainly on two aspects:

- How can such interfaces be specified?

- How can one object or application use another object which implements a known interface?

These issues are shown on the basis of the essential models for fields and uncertain values discussed in chapters 4 and 5. Here, the order of the discussion is reversed, however. Uncertainty representations are dealt with first as they are also used for the field representation. The examples are complete in the sense that the Java programs can be compiled and executed. Many details such as error checking that would be necessary in a real implementation have been left out for simplicity. An understanding of the examples requires the program code to be read[1].

8.2 Uncertainty representation

8.2.1 Overview

Chapter 5 discussed the digital representation of physical quantities. The title of that chapter is "modeling uncertainties", which is more or less synonymous with "representation of physical quantities" because inherent uncertainties are a fundamental characteristic of any measurement. The issues that have to be addressed by a suitable method for the representation of physical quantities can be summarized as follows:

Uncertainties The uncertainty of a value needs to be represented *in the way it was assessed*, e.g., as mean μ and variance σ if those were the parameters resulting from the measurement. This means that the representation does not impose a specific uncertainty modeling method but leaves it open to the concrete case to choose an appropriate one from a predefined set (such as the techniques discussed in chapter 5) or use a custom uncertainty model instead. Possible representation techniques also include methods where the value of the quantity changes with every access such as in Monte Carlo simulations.

Null values Strictly speaking, *null values* are not necessarily a representational issue of a physical quantity but an issue of the *container* of the physical quantity, i.e., a physical measurement is only called a physical measurement if it exists. However, in a computational environment it is often very convenient to be able to represent a physical quantity which does not exist or which is not applicable in the specific context. For example, if there is a time series of measurements and one measurement in the series is missing, then it is useful to assign the missing measurement a *null value* instead of leaving it "empty". In relational databases, for instance, every attribute of a tuple needs to have a value, so that an attribute gets a *null value* if it is missing or undefined. For many applications it is also suitable to differentiate several types of *null values* (Stumptner, 1990). It is beneficial, yet still rather uncommon to distinguish between *missing but applicable* values and *missing because inapplicable* values[2].

---

[1]The examples are available electronically, see Appendix A on page 209.

[2]Most database systems and GIS support only one type of null value and it is up to the user to interpret a null value. This leads sometimes to the need to construct application-specific structures, for instance, for the encoding of completeness (Brassel *et al.*, 1995).

Units and dimensions Most physical quantities X are a product xu of a number x and a unit u as was discussed in section 5.6 on page 112. Thus, a digital representation of physical quantities needs to consider *both* the number x and the unit u in order to capture the measurement's semantics as closely. The inclusion of units has already been discussed (and criticized) in the context of the evaluation of a new programming language for the US Department of Defense in the seventies[3] (Dijkstra, 1978). Presently, there is a proposal for an extension to Java for the support of dimensioned numbers (van Delft, 1996).

Mathematical expressions A representational form needs to support the evaluation of mathematical expressions. This includes arithmetic operators $\diamond \in \{+, -, \times, \div\}$ and other standard functions such as trigonometric functions.

Symbolic computation Symbolic computation would be very useful. On the one hand, error propagation algorithms can make use of symbolic differentiation; on the other hand, the management of units would benefit from such capabilities. However, symbolic computation is still available only within specialized software packages such as *Mathematica* (Wolfram, 1988).

Numerical overflows The values used to represent a measurement need to be able to represent overflows in their digital representation. This means that the largest positive and negative representable value must be defined. Moreover, zero needs to be *signed* in order to distinguish underflows from positive and negative numbers. Negative zero compares equal to positive zero; they can be distinguished, however, by dividing any positive number n by them, i.e., $n/+0 = +\infty$ whereas $n/-0 = -\infty$. Finally, representation of "invalid" numbers (often called *NaN* for *not-a-number*) such as the result of a division by zero is required. These special number representations are provided by the IEEE 754 floating point format (IEEE, 1987) and available in Java with the standard `double` and `float` types.

A specification model based on these requirements and the essential model discussed in chapter 5 is shown in figure 8.1.

8.2.2 A Java package for uncertain values: vds.value

The Java package `vds.value` (appendix A.2.1) implements some of the interfaces and classes from figure 8.1. The root[4] of the hierarchy describing uncertain values is the class `Value` (example E8). It defines a state which is

---

[3] Which eventually became *Ada*.
[4] It is actually the root of this *sub*-hierarchy. In Java, everything inherits from `Object` (see listing 7.2 on page 163), i.e., all classes are contained in the same hierarchy.

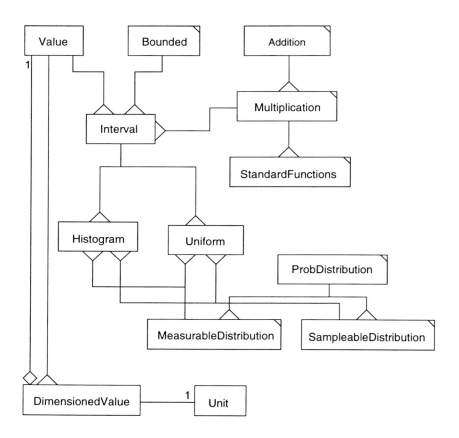

Figure 8.1: Value classes and interfaces

able to distinguish between various *null* values. All other value types inherit from this class.

This package contains also three fundamental interfaces, Addition (example E2), Multiplication (E3), and StandardFunctions (E4). These interfaces declare the basic arithmetics of values of any type and a set of standard mathematical functions such as trigonometric functions. The Addition interface defines methods for the addition $a+b$ and subtraction $a-b$ of a value b to the current object a and also for the determination of its additive inverse $-a$ and the neutral element 0. Similarly, the Multiplication interface extends Addition and defines ab, a/b, a^{-1} and 1 as neutral element. At this stage, a few drawbacks of the current definition of the language Java can be seen in these examples:

- As there is no operator overloading available in Java such as in C++, a sum $a + b$ has to be written as a.add(b). Some expressions become rather complicated (e.g., $(a+b)/(ab)$ is (a.add(b)).div(a.mul(b))). Sometimes, however, operator overloading also hides too much of the underlying complexity and it is not necessarily bad that one is aware of what method invocations are really occurring when evaluating an expression. This means that having to write expressions in this rather unnatural manner seems not to be a great disadvantage, and the gain from refraining from operator overloading is more important than the loss in readability.

- There are not yet any parameterized types in Java. These interfaces fully define the return type of every method call. This means, for example, that the method add() in interface Addition always returns an object of type Addition, i.e., any object that implements this interface. This is, on the one hand, very useful as the applications using objects through such an interface know exactly what to expect. However, this leads often to the necessity of cumbersome *down-casting* the reference to a proper data type within objects implementing that interface.

A fourth interface called Bounded (E1) defines four methods for any bounded value A. The inf() and sup() define the lower and upper bounds of that value, i.e., $x \in A \rightarrow \inf A \leq x \leq \sup A$. Note that this is *not* equivalent to the definition of an interval. For an interval I, it additionally holds that $\inf I \leq x \leq \sup I \rightarrow x \in I$, that is, an interval contains *all* points between the bounds. We do not imply this by the interface Bounded. However, an interval is a very natural example of a bounded value.

The following sections show a few implementations based on these interfaces as well as examples which use the implementations. The implementations and examples are by no means comprehensive. Rather, they are meant to illustrate the concepts derived here. Many topics such as rigorous error checking have been omitted for the sake of simplicity.

8.2.3 Intervals

The class Interval (example E9) is an implementation of the interfaces Bounded and Multiplication. The arithmetic rules are the standard interval arithmetics as given in (Mayer, 1989). For the sake of brevity, the standard functions as defined in interface StandardFunctions are not implemented in class Interval. The features of this value representation are shown in the example class IntervalTest (example E25). This is a stand-alone application which creates instances of Interval and shows their properties. The output of IntervalTest is shown in listing 8.1.

Note that the class Interval does not check for missing or inapplicable values when performing the arithmetic operations. In a real implementation, these methods would return the appropriate null value if the actual instance or the parameter object is tagged as a null value, e.g.:

```
public Addition add(Addition a)
  throws IncompatibleTypesException {
  if ( this.isMissing() || a.isMissing() )
    return Value.Missing();
  if ( this.isNotAppliable() || a.NotApplicable() )
    return Value.NotApplicable();
  ...
}
```

In that way a user of a VDS which contains null values need not consider these special cases. Null values are automatically propagated through all operations.

These examples show also some drawbacks of such an implementation if only single inheritance is allowed. For example, the arithmetic rule for subtraction can always be derived by addition with the additive inverse, i.e., $a - b = a + (-b)$. Therefore, it would be useful if such a rule could be specified once and inherited from all classes implementing the interface. However, a fundamental property of an interface in Java is that it does not carry any implementation. While this is very useful in many cases and even a prerequisite for some of the examples shown here, it forces the class Interval to provide an implementation of the subtraction itself. Sometimes this too can be an advantage. The class Interval does not actually call the addition method with the inverse of the argument (add(a.neg())) but provides a full implementation of the subtraction, and as such, better performance.

8.2.4 Probability distributions

Probability distributions have been discussed as a major way of representing uncertain values in chapter 5. The example here is based on three interfaces, ProbDistribution (example E5), MeasurableDistribution (example E6) and SampleableDistribution (example E7). ProbDistribution provides a

Listing 8.1 Output of `IntervalTest`

```
Interval Test
-------------
a          [1.9,2.1]
b          [4.2,4.4]
c          5.1
a center   2
a radius   0.1
a+b        [6.1,6.5]
a-b        [-2.5,-2.1]
a*b        [7.98,9.24]
a/b        [0.431818,0.5]
a+c        [7,7.2]
a-c        [-3.2,-3]
a*c        [9.69,10.71]
a/c        [0.372549,0.411765]
0          [0,0]
1          [1,1]
```

generic interface to any value which can be modeled as a random variable X with a certain probability distribution $p(x \leq X)$. It contains (accessor) methods for basic properties such as the mean[5] $\mu = \widehat{\mathcal{E}[X]}$, the standard deviation $\sigma = \widehat{\mathcal{V}[X]}$ and the median m with $P(m \leq X) = 0.5$.

`MeasurableDistribution` is an extension of `ProbDistribution` and defines two methods to determine the (cumulative) probability $P(x \leq X)$ and quantiles $q(p) : p = P(x \leq q)$ of a distribution. Generally, `MeasurableDistribution` can be only implemented by distribution types where the cdf and its inverse are known. For normal distributions, for instance, there is no analytical way of describing the cdf because the probability density is a function of the form e^{-x^2} and cannot be integrated analytically. However, there are good numerical approximations of the cdf of a normal distribution and its inverse which can be used in a concrete application.

`SampleableDistribution` extends `ProbDistribution` and provides two methods that can be used to generate random samples which are distributed according to the respective distribution. Note that `SampleableDistribution` is not an extension of `MeasurableDistribution`, even though random samples can be easily derived as soon as there is a quantile function available[6].

---

[5]The values here always represent *measured* entities. Therefore, we use *mean* and *standard deviation* instead of distribution properties such as *expectation value* and *variance*.

[6]The quantile function as the inverse of the cdf maps values p in the range $[0, 1]$ to the corresponding values q with $p = P(x \leq q)$. A random number generator creating uniformly distributed samples in the interval $[0, 1]$ can therefore be used to create samples that follow a particular distribution using the quantile function. Most pseudo-random number generators available produce uniformly distributed values and can be used therefore for the generation.

The reason for the inheritance directly from `ProbDistribution` is that sometimes there are already pseudo-random number generators available for the specific distribution type, so that it is feasible to implement a `SampleableDistribution` without having to provide the cdf and its inverse as required by the interface `MeasurableDistribution`. For normal distributions for example, there are several pseudo-random number generator algorithms that produce normally distributed samples without numerical integration of the cdf. `MeasurableDistribution` provides *two* methods, which are called `newSample()` and `getSample()`. This is required because sometimes a certain sample value is needed more than once and it makes sense to provide storage for the sample within the respective class. Consider an expression of the form $a + b(a + c)$. In a Monte Carlo simulation we need to refer twice to the same sample of a in every run.

Here, the class `Uniform` (example E10) is an example of such a random variable representation. It implements both the interfaces `MeasurableDistribution` and `SampleableDistribution` and is an extension to `Interval`. This means, that it inherits all the arithmetic rules defined by `Interval` on the expense of propagating `Uniform` values to `Interval` values when doing arithmetic operations. A more comprehensive implementation could provide specific arithmetic rules for `Uniform` which can be derived because a uniform cdf can be integrated. Note, that this class does not provide any means of defining covariance structures with respect to other values. It is still an open issue how best to represent covariances in such a model. One way would be to provide the ability for any value to hold a list of associated values with non-zero correlation. However, in many cases the covariance structures are generalized into global characteristics such as an autocorrelation function. It would not be feasible, for instance, to store individually the explicit correlation for every value z_i with all neighboring values z_j. In such cases, the values could contain a reference to one or more central "covariance structures" which can then also be used to provide, for instance, improved spatial interpolation.

The example program `UniformTest` (example E26) uses the `Uniform` class to instantiate a value with a uniform distribution and shows some of the properties of class `Uniform`. The output of `UniformTest` is shown in listing 8.2.

8.2.5 Histograms

Histograms are included here as another kind of representing (empirical) distributions. The class `Histogram` (example E11) defines a histogram and has the same inheritance as the class `Uniform` discussed above. `Histogram` stores absolute frequencies h_i in a given range $[x_0, x_n]$ and a given number n of classes of equal width. `Histogram` provides a method `addValue()` which can be used to dynamically add new samples to the histogram. As such, `Histogram` can be used as a "sink" in a Monte Carlo simulation that aggregates results from every iteration. The relative frequency $r_i = h_i / \sum h_i$ is interpreted as a constant probability in the interval $[x_i, x_{i+1}]$, i.e., the proba-

Listing 8.2 Output of UniformTest

```
Uniform Test
-------------
a                   U[0,2]
x                   1.5
a center            1
a radius            1
mean                1
stddev              0.57735
median              1
p(a<x)=             0.75
x w/ p(a<x)=0.9     1.8
```

bility density is given by $r_i/(x_{i+1} - x_i)$. Using this probability interpretation Histogram is able to provide all the methods from MeasurableDistribution and SampleableDistribution. This means that Histogram can be used as a *source* for Monte Carlo simulations as well. This is very useful if there are several distinct areas in a model where Monte Carlo simulations are to be performed and those areas cannot be combined into a single iteration step.

The example program HistogramTest (example E27) defines two uniformly distributed random variables a and b and determines the product $c = ab$ using a small Monte Carlo simulation. The output of HistogramTest is shown in listing 8.3.

Listing 8.3 Output of HistogramTest

```
Histogram Test
-------------
a                   U[0,2]
b                   U[4,6]
x                   1.5
c mean              5.03475
c stddev            3.09769
c median            5.01087
p(c<x)=             0.165
x w/ p(c<x)=0.9     9.32927
```

A second example, HistogramGraphTest (example E28), shows another interesting possibility of a distributed implementation based on Java. The package vds.value which defines the classes and interfaces discussed here could also contain some classes which provide simple visualization capabilities for some of the value classes. This is possible since the Java runtime environment contains the so-called *Abstract Windowing Toolkit (AWT)*. This means that a VDS defined on one host could contain classes which provide

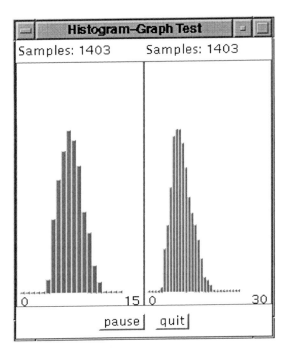

Figure 8.2: Example window of HistogramGraphTest, X Windows

visualization and which can be used by the data user, even if the data user is running a different operating system with a different windowing system. Consider for example the class HistogramPane (example E12). This class defines a GUI element that produces a visualization of an associated histogram. A remote application could now use this class to display a histogram within its own windowing system, i.e., the displaying capabilities are defined on a different platform and operating system, but can still be used by a VDS user[7]. The example program HistogramGraphTest (example E28) uses the class HistogramPane to display two animated histograms of a running Monte Carlo simulation calculating $c = ab$ and $d = ac$. a and b are uniformly distributed, c and d are histograms displayed in two HistogramPanes. Note that d is determined using random samples from histogram c. The window of HistogramGraphTest is shown for two different operating systems in figures 8.2 and 8.3.

---

[7] A similar example is shown below for "configurable" fields.

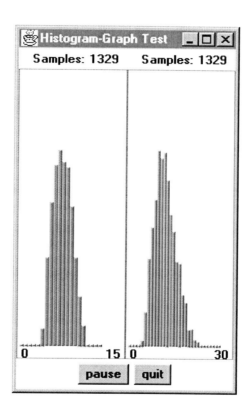

Figure 8.3: Example window of `HistogramGraphTest`, Microsoft Windows 95

8.2.6 Dimensioned values

The last example of the value representations concerns dimensioned numbers. It is based on standard Java and does not need enhancements to the language as proposed in (van Delft, 1996). The basic classes defined here are `Unit` and `DimensionedValue`. `Unit` (example E13) defines a measurement unit. The state of the unit consists of a name of the unit and a list of conversions to other units. Here, we assume ratio type numbers where the conversion is a simple multiplicative factor. The method `addConversion()` allows a new entry to the conversion table[8] to be added. It simultaneously adds a reciprocal factor to the conversion table of the other unit. For ratio numbers only it would not be necessary to have a conversion table per unit, a system-wide conversion matrix would suffice. However, for other value and unit types it might be necessary to have individual conversion functions to other units for *each* unit. The `Unit` class provides a convenience-method `convertTo()` which converts a value from the current unit to a passed unit. Note that `Unit` shows a special feature of Java. An *initializer block* is used to initialize the conversion table between *meter* and *centimeter*. This block is executed when the `Unit` class is loaded for the first time and provides a convenient way to initialize static components. In a real implementation, one could read some configuration file at that point at load time to populate the various conversion tables.

`DimensionedValue` (example E14) uses `Unit` for a specialization of class `Value`. This specialization provides an association of a unit and any value type that can be multiplied (i.e., that implements `Multiplication`). `Dimensioned-Value` contains some logic for the arithmetics of dimensioned numbers. For example, it allows only numbers with equal or compatible units to be added or subtracted. And, when multiplying two *dimensioned numbers* the multiplication rules try to simplify the unit expression by performing a simple dimensional analysis. For example, dividing two numbers of the same unit results in a dimensionless number.

This is shown in example `DimensionTest` (example E29). This class defines four dimensioned values. All values are represented by intervals and have a unit. The output of `DimensionTest` is shown in listing 8.4 and demonstrates the dimensional analysis. Note that the last expression in `DimensionTest` tries to add a meter value to a Newton-square value and accordingly raises an exception. Note also that `DimensionedValue` works with *any* value type derived from the "root" class `Value` which implements `Multiplication`, i.e., not only with intervals as is shown in the example. A concrete implementation would, however, need more sophisticated rules for unit conversions and dimensional analysis. The example here only demonstrates that a simple dimensional analysis can be easily achieved.

---

[8] The conversion table is maintained as a hash table with a unique hash code per unit.

Listing 8.4 Output of `DimensionTest`

```
Dimension Test
--------------
a           [1.9,2.1] m
b           [102,104] cm
c           [6.1,6.5] N2
d           [1.1,1.2] N
a+b         [2.92,3.14] m
a-b         [0.86,1.08] m
a*b         [1.938,2.184] m2
a/b         [1.82692,2.05882]
c/d         [5.08333,5.90909] N
(c*a)/(b^2) [0.107156,0.1312] N2cm-1
error vds.value.IncompatibleTypesException: m N
```

8.3 Fields

8.3.1 Overview

The examples of field implementations are based on the OGIS coverage model shown in figure 3.3 on page 52 and detailed in (Gardels, 1996), as well as on the theory developed in chapter 4. The implementation of a field has to represent the elements of equation 4.1 on page 66, i.e., a domain $\mathbb{D}$, a range $\mathbb{V}$ and a mapping $s \rightarrow z(s)$ from $\mathbb{D}$ to $\mathbb{V}$. The model of the example implementation discussed here is shown in figure 8.4 and is based on an interface `Field` that offers these basic field attributes. Interface `Field` (example E23) defines a generic scalar field on a two-dimensional geometry. The restriction to a *scalar* field and a *two-dimensional* geometry has been made for the sake of simplicity only and does not make a fundamental difference.

The `getDomain()` method of the interface `Field` returns the domain $\mathbb{D}$ as a `Geometry` object. In order to keep the examples simple three geometry classes have been defined in a package `vds.geometry`. These geometry classes define points (class `Point` shown in example E20), rectangles with sides parallel to the coordinate axes (class `Rectangle`, example E21) and polygons (class `Polygon`, example E22). `Geometry` itself is not a class but an interface defining a method which determines if a point s lies inside the object and a method that derives the minimum bounding box (parallel to the coordinate axes) of the object (see example E19). Both `Rectangle` and `Polygon` implement this interface. The object returned by the `getRange()` method of a field can therefore be mainly used to determine whether a point lies in there. Note that there is no notion of spatial reference system, projection or scale. These are issues that have to be dealt with on a more general level in a geoprocessing framework since these issues affect *all* spatially referenced entities. OGIS addresses this issue by distinguishing between *location* and *geometry* as has

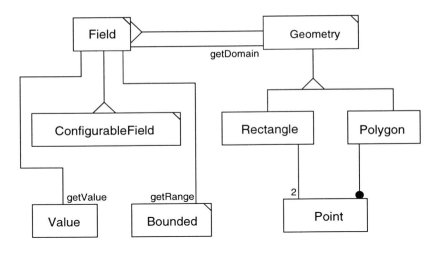

Figure 8.4: Field and geomtery classes and interfaces

been discussed earlier in chapter 3.

A second accessor method `getRange()` of interface `Field` retrieves the range $\mathbb{V}$. Since `Field` represents a *real-valued scalar* field the return type of `getRange()` is any `Bounded()` object, i.e., something which has upper and lower bounds.

The third method of `Field` is the central one as it defines the mapping $s \to z(s)$. The method `getValue()` accepts a `Point` as argument and returns a `Value`. This means that the return object can be *anything* that inherits from `Value`, e.g., all the various uncertainty representations discussed before. The spatial reference passed here is always a point. One could imagine an extension with more advanced analysis methods such as:

```
public interface AggregateableField extends Field {

    // inherited from Field:
    // public Value getValue(Point p);

    // the average over Geometry g
    public Value getAverage(Geometry g);

    // the maximum over Geometry g
    public Value getMaximum(Geometry g);

    // ... etc.
}
```

The next two sections show sample implementations of a VDS based on these interfaces. The first example creates a simple dataset representing a field with irregular sampling points. The second example extends this towards tight integration into graphical user interfaces.

8.3.2 A simple field

Class `MyField` (example E30) is an implementation of interface `Field`. It represents a real valued scalar field (as required by `Field`) which is based on irregularly distributed sampling points $\{s_i\}_1^n$ with $n = 3$. The *state* of this VDS is directly contained in the code, i.e., there is no other data item needed than the Java class itself. In a real implementation, however, one would typically retrieve the state from an associated data stream (file or network connection) or a database query.

The constructor of `MyField` creates the three sampling points s_1, s_2, s_3 and associates them in a *relation* with the corresponding field values. The relation is maintained in a hash table using the point reference as key. The field values are represented by intervals (class `Interval`). The domain $\mathbb{D}$ is defined as a polygon with 5 nodes.

The most interesting method of `MyField` is the implementation of the method `getValue()`. This method first checks whether the passed point s lies inside the domain $\mathbb{D}$. If $s \notin \mathbb{D}$ then the method returns with a "not applicable" value as defined in class `Value`[9]. Otherwise, a field value at s is determined by a simple distance-weighing interpolation (which is feasible here with only 3 points). Note that the interpolated value is of type `Interval` as well and the calculation uses interval arithmetics, i.e., preserving uncertainties present in the sample values.

The VDS represented by `MyField` is used in a sample application `FieldTest` (example E31). Class `FieldTest` first displays some of the field's properties and then retrieves and prints the field values on a regular grid. This is exactly what would be needed, for example, to visualize the field data on a computer display. The output of `FieldTest` is shown in listing 8.5. Note that `FieldTest` does not need to know what value type is returned by method `getValue()` of class `MyField()`. The implicit conversion to a `String` calls the appropriate `toString()` method, which displays here either the interval bounds (if it is an `Interval`) or "n/app." if it is a "not applicable" value.

8.3.3 VDS and user interfaces

The last example `GUIFieldTest` (see E35) combines the issues discussed before and adds two interesting features which are based on Java's multi-platform ability. First, it defines a new VDS which is based on a specialized

---

[9]One could also imagine an exception being raised for such a case. However, that would probably make the use of such a VDS tedious, because one would have to manage these exceptions after every call to `getValue()`. Using a *null-value* representation allows the calling application basically to ignore the possibility of missing or inapplicable values as they are represented as special value types.

Listing 8.5 Output of FieldTest

```
Field Test
----------
Domain  vds.geometry.Polygon[
  points=5,
  MBR=vds.geometry.Rectangle[x=0.5,y=0.5,width=8,height=6.5]]
]
Range   [0,6]
MBR     vds.geometry.Rectangle[x=0.5,y=0.5,width=8,height=6.5]
vds.geometry.Point[x=1,y=1]   = [2.50154,2.75126]
vds.geometry.Point[x=1,y=2.5] = [2.10204,2.30224]
vds.geometry.Point[x=1,y=4]   = [2.46455,2.68365]
vds.geometry.Point[x=1,y=5.5] = [2.86565,3.08694]
vds.geometry.Point[x=1,y=7]   = n/app.
vds.geometry.Point[x=2.5,y=1]   = [3.20159,3.55434]
vds.geometry.Point[x=2.5,y=2.5] = [2.57098,2.81989]
vds.geometry.Point[x=2.5,y=4]   = [2.83225,3.05718]
vds.geometry.Point[x=2.5,y=5.5] = [3.07941,3.28267]
vds.geometry.Point[x=2.5,y=7]   = n/app.
vds.geometry.Point[x=4,y=1]   = n/app.
vds.geometry.Point[x=4,y=2.5] = [3.23803,3.56994]
vds.geometry.Point[x=4,y=4]   = [3.04221,3.28274]
vds.geometry.Point[x=4,y=5.5] = [3.08254,3.29029]
vds.geometry.Point[x=4,y=7]   = n/app.
vds.geometry.Point[x=5.5,y=1]   = n/app.
vds.geometry.Point[x=5.5,y=2.5] = [3.41285,3.76896]
vds.geometry.Point[x=5.5,y=4]   = [3.15217,3.42835]
vds.geometry.Point[x=5.5,y=5.5] = [3.08702,3.32702]
vds.geometry.Point[x=5.5,y=7]   = n/app.
vds.geometry.Point[x=7,y=1]   = n/app.
vds.geometry.Point[x=7,y=2.5] = [3.3976,3.74818]
vds.geometry.Point[x=7,y=4]   = [3.22128,3.52161]
vds.geometry.Point[x=7,y=5.5] = [3.13291,3.40223]
vds.geometry.Point[x=7,y=7]   = n/app.
```

interface ConfigurableField. ConfigurableField (example E24) is an extension of Field and adds a single method configure(). This method allows a VDS to configure itself to specific user needs *with a graphical user interface*. This means that a dataset is able to display a dialog window on the application's behalf. This is much like printer configuration controls seen in some operating systems. There, the dialog that lets the user specify some special, printer-specific settings is usually displayed and managed by the printer driver program. An application can request the display of the dialog from a printer driver. The difference here with respect to printer drivers[10] is, however, that the printer driver is typically installed on the local machine and is written for the local system architecture and operating system. Classes (or VDS) based on ConfigurableField can display their configuration dialog(s) on whatever system the VDS is executing since Java does provide a platform-independent access to graphical user interfaces through the Abstract Windowing Toolkit (AWT).

Class MyConfigurableWindow (example E32) is an example of an implementation of interface ConfigurableWindow. The sample class MyConfigurableWindow extends MyWindow and adds the method configure() as required by interface ConfigurableWindow. This method displays a modal dialog that lets the user change some settings such as, for instance, the interpolation method to be used or some magic parameters. The configure() method gets a context object passed. This allows a VDS to store user-specific settings in that context object. The context object itself is managed by the calling application. This is especially useful if multiple applications are accessing a single VDS in a client/server model.

The second feature introduced in GUIFieldTest is a special *class loader*. This class loader is implemented in class FileClassLoader (example E34) and demonstrates how a VDS can be instantiated from any data source. Here, a VDS is loaded from the local file system, but it is straightforward to extend method loadClassData() of FileClassLoader to load a VDS from a network stream or a database query. The application GUIFieldTest can present a file selection dialog (see figure 8.5) which lets the user select a VDS to load. Using the class FileClassLoader the file name (which is a String object) can be turned, so to speak, into a running instance of the corresponding class. Also, the class loaded can be *any* class that implements class Field. Using the runtime type information available in Java GUIFieldTest can determine whether a class implements a certain interface and whether it is an instance[11] of a class or any of its subclasses. GUIFieldTest uses this to check if the VDS loaded does really represent a field and if it is a configurable field. If so, the corresponding menu item is enabled which lets the user invoke the dataset's

---

[10] Other software components sometimes similarly interacting via graphical user interfaces are drivers for scanners and database middle-ware.

[11] Java uses an instanceof operator for this. The same operator is, by the way, used to determine if a class implements a certain interface, i.e., if myclass instanceof myinterface is true, then myclass does implement myinterface.

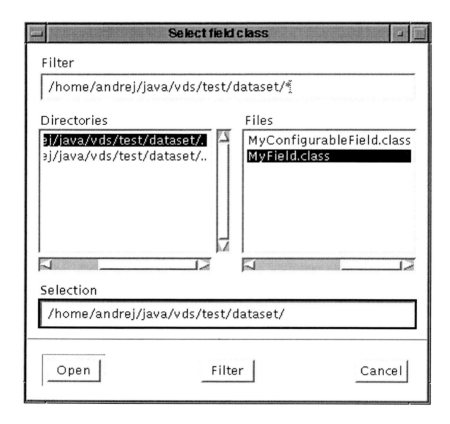

Figure 8.5: Specifying a VDS to load

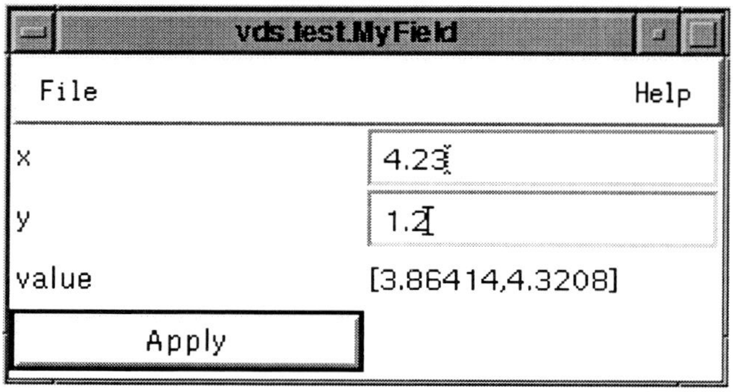

Figure 8.6: Example window of `GUIFieldTest`

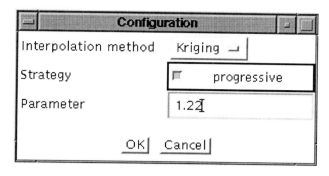

Figure 8.7: Dataset with a configuration dialog

own configuration dialog (which is, again, part of the dataset loaded).

The main window of GUIFieldTest is shown in figure 8.6. It offers two input text fields where coordinates can be entered. If the button "apply" is pressed the field value of the currently loaded field at the entered coordinates is displayed (i.e., the field's getValue() method is called). Figure 8.7 shows the dialog that pops up when loading VDS MyConfigurableField and selecting the menu item *configure*.

CHAPTER NINE

Conclusion

9.1 Retrospective view

Integration and interoperability issues in geoprocessing have been discussed on various abstraction levels in the previous chapters, ranging from general overview to conceptual modeling and implementation considerations. At the very heart of all these discussions is the *common contract* that interoperating parties need to agree on. The main optimization problem that needs to be addressed there is the trade-off between *rigid specification* and *flexibility*. On the one hand, rigid specification is necessary in order to provide a good basis for interoperability. On the other hand, flexibility and ease of use is needed to achieve acceptance by the interoperating parties. Data sharing strategies try to make a reasonable compromise within this area of tension, a compromise which is dependent on the data sharing context. Within well-defined information communities such as organizations and enterprises it is often feasible to adopt rather rigid protocol specifications, because there are structures in place that help the implementation of such regulations[1]. In other words, there is not necessarily much autonomy for the interoperating parties. Data sharing partners possessing a certain autonomy usually prefer approaches with more flexibility at the expense of specification.

The objective addressed in this thesis is a discussion of methods to define and implement such contracts or protocols between interoperating parties, trying to find a better compromise between the two poles and going beyond traditional approaches based on, for example, file format standardization. The

---

[1]These interoperability issues could be called "intraoperability" similar to the notion of "intranet" which describes the usage of TCP/IP-based technologies known from the Internet within enterprise-wide networks.

analysis of the interoperability requirements and impediments has identified semantical heterogeneities as a core problem in the context of geoprocessing. These heterogeneities need to be overcome when sharing and exchanging data and services[2], that is, the contract defined needs to address these heterogeneities and provide means to cope with or overcome them. The approach proposed here is called *Virtual Data Set* and addresses heterogeneities by adopting an object-oriented model. An object-oriented model allows a compromise which increases both the degree of specification and the flexibility:

- Object-oriented models are much more *expressive* than, for instance, static data descriptions as it is possible to define methods to model the object's behavior.

- Encapsulation allows implementation details to be hidden from the object user, reducing the overall complexity.

- The collection of methods and attributes of an object or a class, respectively, is a way to provide a well-defined specification.

- Object-oriented approaches allow a more intuitive way to model reality.

Therefore, object-oriented models can reduce semantic heterogeneities by providing an environment that allows a semantically richer representation of data and services. For example, a VDS can contain methods which provide various *views* of a dataset corresponding to the requirements of a VDS user. Continuous fields have been chosen as a good example to demonstrate the flexibility of the approach. Data representing continuous fields are frequently needed in forms different from the "raw" data values. A field represented by a VDS provides various views of itself if the VDS has been programmed accordingly.

The design and specification of an interoperability contract based on an object-oriented model needs a well-defined understanding of the "things" that are represented. Here, the "things" are real-world phenomena and thus a thorough analysis of the characteristics of the real-world phenomena with respect to their commonalities and individualities is needed. Using terminology from object-oriented design, *essential models* are required. It is particularly important when the interoperating parties belong to different information communities. Each information community uses its own implicit rules and understanding of the real-world and as such, creates semantic heterogeneities. An object-oriented model needs to capture as much of the semantics of real-world phenomena as possible in order to reduce those heterogeneities. Essential models for two areas have been presented here. On the one hand, *continuous fields* as a major information type in environmental sciences have

---

[2] As mentioned before, we do not differentiate here between *sharing* and *exchange*, or between *data* and *services*. Everything finally results in exchanging pieces of information.

been analyzed. On the other hand, *physical quantities* and their *uncertainties* have been studied to provide an essential model for the representation of physical or uncertain quantities, respectively, in an object-oriented model.

Interoperability also means *distributed* systems. Distributed meaning here that data producers and data users do not necessarily use the same computing infrastructure. Due to the growing use of local and global area networks in the past few years interoperability issues have also started to mean distributed system in the stricter sense.

That is, interoperability is what is needed to produce an integrated computing facility on a collection of autonomous computers linked by a network[3]. Network bandwidth has started to become the dominant growing factor at the time of writing. CPU speed, secondary storage size, storage access time and similar factors that were showing the largest growing rates in the past are starting to be passed by the growth in network speed and accessibility. Most often, data are not exchanged any more by using transportable tertiary storage media such as magnetic tapes or diskettes but by means of network transport protocols such as HTTP or FTP. Not only is the transfer time dramatically decreased (a few minutes versus a few days or weeks), but also the *size* of the datasets being transferred can be reduced since only the required part of a larger dataset needs to be down-loaded.

Thus, the implementation of an object-oriented "contract" can be based on the assumption of a distributed system. Various models can be imagined for the implementation of objects (i.e., VDS) within such a distributed system. The object implementation can be executed at the data producer's site (CORBA), at the data user's site (COM), or any combination of both. Systems such as Java combined with CORBA allow flexible distribution of an object implementation between the data producer ("server") and the data user ("client"). Certain methods can be transferred to the client for local execution, other methods can be executed on the server, for example, if some special high-performance hardware available on the server is needed to execute the method.

The discussion of the properties of various software systems that can be used for the implementation of such a distributed object system has shown that Java possesses a series of properties which make it a good candidate for the implementation of concepts and specifications such as VDS and OGIS. The examples in chapter 8 demonstrate these concepts with a some examples. These examples, however, were deliberately kept simple and illustrative. A *real* implementation needs to be based on a comprehensive specification such as OGIS, defining essential models for other core elements of geoprocessing as well. For example, geometry issues, spatial reference systems and similar topics have been completely left out here. However, the examples give strong evidence for the viability of the concept, even if the elegance of the implementation is very much dependent on the choice of Java. Using

---

[3]This is an inversion of Coulouris, Dollimore and Kindberg's definition of a distributed system [p. 1](Coulouris *et al.*, 1994).

C++, for example, would have made certain examples extremely tedious or even impossible. But, Java was also intentionally chosen because it shows promising characteristics which would also make it a reasonable choice for real, production-quality software development and as such, for the implementation of OGIS. The most important benefit from using Java is that it allows the transfer of *executable* contents from data producer to data user, which means that an object implementation prepared by the data producer can be used by the data user on whatever platform. Therefore, data exchange can profit from an existing network and a distributed environment, but *it is not limited to* these. The executable contents can also be transferred by traditional tertiary media. Even if network bandwidth continues to grow in the near future there are always situations where a dataset needs to be used on a system which is not connected to the data producer by any network, for example, due to some security considerations.

Many of the themes discussed in this thesis are not new as such. Distributed systems have been around for a while, so have object-oriented concepts. The combination and *integration* of these various issues, however, can be seen as the major scientific contribution of this thesis. The *Meta*-integration provides strong arguments for a geoprocessing infrastructure based on a three-tier model as proposed here by the VDS concept. A thesis about integration must necessariliy have a wide scope, making generalizations where possible but retaining particularities where generalizations are not possible. In science, there is always a trend towards more specialization, "increasing" the knowledge of human kind by roughly a million scientific publications a year. Methodological sciences such as applied computer science are somehow orthogonal to these specializations in that the overall complexity of hard- and software systems calls for reusable solutions. That is, a computing infrastructure needs to be as general as possible. But still, it has to solve the domain scientist's specialized computing problems. The classical trade-off here is between the extremes of a system which has zero features and is useful for everyone and a system which has inifinitely many features and is useful for no one. The bottom line of this thesis is that an optimum between these two extremes can be approached by a *decomposition* into a set of interoperable software components. Each software component provides a part of the overall system, be it a dataset, a special algorithm, facilities for human/computer interaction and so on. The *interaction* between the components happens through well-defined interfaces. These interfaces are clearly the *key* element within such an architecture. If they are not expressive enough, the overall system remains specialized. If they are too complicated, no components would be created whatsoever. If they are proprietary, vendor independence is lost. That's why the essential models are so important. The essential models provide the basis for the interface specification. If the essential models capture a model of reality on which many people agree then the derived interfaces will be useful for them.

In this thesis, two problem domains have been selected to demonstrate the role of essential models. *Continuous fields* and *representation of physical quantities* are fundamental parts of a spatial information theory and, thus, also fundamental for a geoprocessing infrastructure. Following the generalization principle these essential models try to *minimize* the number of concepts needed to describe continuous fields and physical quantities. These two essential models – being "only" exampes here – can be seen themselves as a scientific contribution to a spatial information theory in that a clear theoretical foundation is given.

This work shares the same overall objectives as the OGIS project. These general objectives, however, go far beyond what was reached here and in the current OGIS efforts. The aspects adressed here are to some degree complementary to the more problem-driven context of OGIS in that they provide the *metadiscussion* of interoperability topics while OGIS is mainly concerned to find the optimal specification for that "well-defined interface" mentioned above. This thesis explains in part I why it makes sense to do that. Part II discusses the *foundation* of such a specficiation by virtue of two examples. Part III finally discusses architectural and implementation aspects, but – and this is important to note – *not on a level of an implementation specification. This thesis does not provide a specification.* Specification is addressed by OGIS and it is addressed there very well.

9.2 Prospective view

Some of the ideas presented in this thesis are not very new. Nor are Java's features revolutionary. Java is a combination of concepts that have been around for a while, and every C++ programmer was wondering when Java came out why it did not happen earlier. This applies also to the concept of VDS. Many system components that we use in day-to-day computing show properties of a VDS, even though these components are often parts of the operating system and usually not visible. In fact, an operating system itself is a good model for interoperable systems if we consider its various parts as autonomous parts communicating which each other. The individual components need to be autonomous because they might stem from different sources. For example, a component which is a driver for a certain piece of hardware might be produced by the hardware vendor itself. It needs to comply to the interfacing specifications for the operating system under consideration but is more or less free concerning the internal details. Thus, component-based systems are already in place today, but usually in a small, well-defined and well-controlled context such as an operating system.

VDS, OGIS and similar approaches promote the adoption of component-based systems in a larger computing context. A geoprocessing environment can be imagined as a collection of interoperating services. These services might represent datasets (data sources), agents performing certain tasks on a user's behalf, analysis modules, visualization tools (data sinks), measurement devices and anything else which can be identified as a well-defined component.

There is no difference between datasets and their processing. A dataset is as much a service as any analytical model. Information flows from datasets to other services much like the Unix piping model. The most important benefit of such a structure is the reduced complexity of the overall system. For example, a visualization component probably has a limited but specialized functionality, i.e., it does a few things but does them very well. The component does not bother about particularities of various data sources as long as these data sources conform to the common contract. Thus, the visualization component can display on-line data from a real-time measurement (e.g., a flight-control radar), data produced by a simulation model component, data from a test data generator or simply the normal case of data derived from static samples of the real world (e.g., a map).

From a data user's perspective it seems to be a very promising future. Everything is "plug-and-play", there are no more tedious and error-prone conversions, transformations and the like. One can assemble the best system ever, buying only the components needed – the *best* components – instead of making a choice between a few gigantic systems which do not meet all the user's requirements but offer too much functionality. This tendency can be seen within all areas of computing, and users are happy with it.

From a data producer's perspective, however, life seems to get more complicated. Instead of producing a dataset in a static format the data producer would be required to produce a software component adhering to the interfaces specified. In practice, a dataset needs to be "programmed". At first glance, this seems to be an unacceptable impediment, making data production very expensive. But, efficiency here is only a question of the tools used. For example, nowadays only a few people program directly in POSTSCRIPT. Most POSTSCRIPT-program code is automatically generated by application programs or operating system drivers. Text processing applications and graphics packages can be used to interactively create graphical content, which is then translated into a POSTSCRIPT program. The application or – in most cases – the printer driver is, so to say, a POSTSCRIPT code generator. The possibility to manually edit and produce POSTSCRIPT code in some special cases is one of the strengths of the POSTSCRIPT concept. A data producer would need the same support for the creation of VDS. Ideally, the data producer's "data manufacturing application" would have a menu item "Save as ..." which would generate VDS much like writing a physical file. The creation of a VDS could also be supported by appropriate interactive applications. For example, techniques known from *visual programming* seem to be very promising here. Visual programming uses "visual expressions such as diagrams, free-hand sketches, icons or even graphical manipulations (...) for GUI programming, depicting relationships and data structure behavior, and visually combining textually programmed units to build new programs" (Burnett & McIntyre, 1995). It is especially useful when a library of classes with well-defined interfaces is available (e.g., OGIS's well-known structures) which need to be combined in some way to create a new program. Visual programming would certainly allow data producers to efficiently assemble a VDS from a given set of utility classes. Specific applications would perhaps still involve manual programming to cover cases which are not directly supported by the visual programming environment. Legacy applications and data formats at the data producer's site can be included using *wrapper classes* which provide translation between internal representations and the external well-defined interfaces.

The aspect of *distributed* systems offers new potential both for data users and data producers. A data collection could be made available on a node of the network by the data producer. This would not allow data users to quickly retrieve a dataset much like it is done today with file transfer. However, the service provided by the data producer could exhibit a certain intelligence. For example, a VDS could be down-loaded into the data user's system. This VDS would use database queries against the data producer's site to retrieve its state depending on the user needs. Therefore, only the information needed would be transferred from data producer to data user. The local VDS would still have some methods executed locally (such as, for example, interpolation, unit conversions) where there is no need to contact the data producer's server. For the data producer there is an additional benefit to such a setup. It would be feasible to set up billing based on effective usage. Such accounting schemes are sometimes called *micro-payments* and intensively discussed in the Internet at the time of writing. Micro-payments would allow a data producer to debit each data access on a predefined granularity. Data users would pay only for the data actually used; data producers would have a sales channel which is easy to maintain and attractive due to the low costs.

System vendors are a third party shaping the future of geoprocessing. Similar developments in other areas show that vendors need to give up proprietary solutions. "Being open" is a key selling feature and demanded by customers. Thus, it is not surprising that most major GIS vendors are members of the OGC and promoting more or less the OGIS approach. These developments will lead to more competition, and it is not necessarily clear what impact that will have on the quality of the products. Certainly, small and medium enterprises will have more chance in participating in the market, and it is also likely that *very* small companies can provide some highly specialized components. This development can be seen in the market of reusable software components for GUI developments. For Microsoft Windows for example, there are some standardizations (based on COM) which allow such elements to be created and reused from within many different programming languages. There is a growing market of small and very small companies selling such specialized "controls" or widgets all over the world, being in competition with large software companies which do not necessarily produce better products on the same level of granularity. Of course, small companies could never compete in a market where the granularity is "one full-featured GIS".

Interoperability has become a *research topic* in the last few years as well. There are many issues which need to be addressed on a more fundamental level. Standardization organizations such as OGC vendors are not always able to go deliberately deep into some details and fundamental questions as there are deadlines to meet and always increasing time-to-market objectives. Research can provide more than "just" some input to these standardization organizations. As has been seen, a fundamental part of a specification is a thorough understanding of the "things" under consideration. In geoprocessing, there is still no really unified and accepted theory of what "spatial information" is. There are still many questions open in order to create the "fundamental essential model", even if this goal can only be approached asymptotically and, probably, never fully reached. The discussion of continuous fields and uncertainty models might have contributed something to a better understanding of spatial information, or, a better *explicit* understanding of spatial information. Developing theories often involves making implicit knowledge explicit, introducing typologies and identifying relationships between the entities. In that context, some concepts of this thesis could be seen as a mosaic stone for spatial information theory.

APPENDIX A

Java examples

A.1 Availability

The following Java examples are available on the World-Wide Web at

http://www.geo.unizh.ch/gis/phd_publications/vckovski.

The programs require Java JDK Version 1.0.2 or higher. The examples have been deliberatley left simple to avoid obfuscation of the main concepts. The code has been grouped into four Java packages:

Package	Content
vds.value	Numerical value representation
vds.geometry	Simple geometry classes
vds.field	Two field examples
vds.test	Test programs that utilize the sample VDS

A.2 Package contents

A.2.1 Package vds.value

Example	Class/Interface/Exception
E1	vds.value.Bounded
E2	vds.value.Addition
E3	vds.value.Multiplication
E4	vds.value.StandardFunctions
E5	vds.value.ProbDistribution
E6	vds.value.MeasurableDistribution
E7	vds.value.SampleableDistribution
E8	vds.value.Value
E9	vds.value.Interval
E10	vds.value.Uniform
E11	vds.value.Histogram
E12	vds.value.HistogramPane
E13	vds.value.Unit
E14	vds.value.DimensionedValue
E15	vds.value.IncompatibleTypesException
E16	vds.value.IncompatibleUnits
E17	vds.value.IllegalProbability
E18	vds.value.OutOfBounds

A.2.2 Package vds.geometry

Example	Class/Interface/Exception
E19	vds.geometry.Geometry
E20	vds.geometry.Point
E21	vds.geometry.Rectangle
E22	vds.geometry.Polygon

A.2.3 Package vds.field

Example	Class/Interface/Exception
E23	vds.field.Field
E24	vds.field.ConfigurableField

A.2.4 Package vds.test

Example	Class/Interface/Exception
E25	vds.test.IntervalTest
E26	vds.test.UniformTest
E27	vds.test.HistogramTest
E28	vds.test.HistogramGraphTest
E29	vds.test.DimensionTest
E30	vds.test.MyField
E31	vds.test.FieldTest
E32	vds.test.MyConfigurableField
E33	vds.test.MyConfigureDialog
E34	vds.test.FileClassLoader
E35	vds.test.GUIFieldTest

References

Adler, R. M. 1995. Emerging standards for component software. *Computer*, **28**(3), 68–77.

Adobe. 1985. *Postscript language reference manual*. Adobe Systems, Inc.

Adobe. 1990. *Postscript language reference manual*. Second edn. Reading, Massachusetts: Addison-Wesley.

Adobe. 1996 (June). *Adobe Printing and Systems Technologies*. White paper.

Al-Taha, K., & Barrera, R. 1990. Temporal data and GIS: an overview. *Pages 244–254 of: Proceedings of the GIS/LIS Conference.*

Albrecht, J. 1995. Semantic Net of Universal Elementary GIS Functions. *Pages 235–244 of: Proceedings of the AUTOCARTO 12 Conference.*

Alonso, G., & Abbadi, A. El. 1994. Cooperative modeling in applied geographic research. *International Journal of Intelligent and Cooperative Information Systems*, **3**(1), 82–102.

Anklesaria, F., McCahill, M., Lindner, P., Johnson, D., Torrey, D., & Alberti, B. 1993 (Mar.). *The Internet Gopher Protocol*. Internet RFC 1436. InterNIC.

Aybet, J. 1990. Integrated mapping systems – data conversion and integration. *Mapping Awareness*, **4**(6), 18–23.

Baier, R., & Lempio, F. 1992. Computing Aumann's Integral. *Pages 71–92 of:* Kurzhanski, A. B., & Veliov, V. M. (eds), *Modeling Techniques for Uncertain Systems*. Basel: Birkhäuser.

Bandemer, H., & Gottwald, S. 1993. *Einführung in Fuzzy-Methoden.* Fourth edn. Berlin: Akademieverlag.

Bandemer, H., & Näther, W. 1992. *Fuzzy Data Analysis.* Dordrecht: Kluwer Academic Publishers.

Barlow, R. J. 1989. *Statistics.* New York: John Wiley and Sons.

Bartelme, N. 1995. *Geoinformatik: Modelle, Strukturen, Funktionen.* Berlin: Springer Verlag.

Batini, C., Lenzerini, M., & Navathe, S. B. 1986. A comparative analysis of methodologies for database schema integration. *ACM Computing Surveys*, **18**(4), 323–364.

Bauch, H., Jahn, K.U., Oelschlägel, D., Süsse, H., & Wiebigke, V. 1987. *Intervallmathematik.* Leipzig: BSB Teubner.

Beard, K., & Mackaness, W. 1993. Visual access to data quality in geographic information systems. *Cartographica.*

Beard, K. M., Buttenfield, B., & Clapham, S. B. 1991. *NCGIA research initiative seven: Visualization of spatial data quality.* Tech. rept. 91-26. National Center for Geographic Information and Analysis, Santa Barbara, California.

Berners-Lee, T., & Connolly, D. 1995 (Nov.). *Hypertext Markup Language – 2.0.* Internet RFC 1866. InterNIC.

Berners-Lee, T., Masinter, L., & McCahill, M. 1994 (Dec.). *Uniform Resource Locators.* Internet RFC 1738. InterNIC.

Berners-Lee, T., Fielding, R., & Frystyk, H. 1996 (May). *Hypertext Transfer Protocol – HTTP/1.0.* Internet RFC 1945. InterNIC.

Blott, S., & Včkovski, A. 1995. Accesing geographical metafiles through a database storage system. *In:* (Egenhofer & Herring, 1995).

Booch, G. 1991. *Object Oriented Design with Applications.* Menlo Park, California: The Benjamin/Cummings Publishing Company, Inc.

Borenstein, N., & Freed, N. 1993 (Sept.). *MIME (Multipurpose Internet Mail Extensions) Part One: Mechanisms for Specifying and Describing the Format of Internet Message Bodies.* Internet RFC 1521. InterNIC.

Bowen, J. P., & Hinchey, M. G. 1995. Ten commandments of formal methods. *Computer*, **28**(4), 57–63.

Brassel, K., Bucher, F., Stephan, E.M., & Včkovski, A. 1995. Completeness. *In:* (Guptill & Morrison, 1995).

Bretherton, F. P., & Singley, P. T. 1994. Metadata: a user's view. *In:* (French & Hinterberger, 1994).

Brockschmidt, K. 1994. *Inside OLE 2*. Redmond, Washington: Microsoft Press.

Bucher, F., & Včkovski, A. 1995. Improving the selection of appropriate spatial interpolation methods. *In: A Thoretical Basis for GIS. International Conference, COSIT '95, Semmering, Austria, September 21-23*. Lecture notes in computer science, no. 988. Berlin: Springer Verlag.

Bucher, F., Stephan, E.-M., & Včkovski, A. 1994. Integrated Analysis and Standardization in GIS. *In: Proceedings of the EGIS'94 Conference*.

Buehler, K., & McKee, L. (eds). 1996. *The OpenGIS Guide*. Wayland, Massachusetts: OpenGIS Consortium, Inc. OGIS TC Document 96-001.

Burnett, M. M., & McIntyre, D. W. 1995. Visual programming. *Computer*, **28**(3), 14–16.

Busch, P., Lahti, P. J., & Mittelstaedt, P. 1991. *The Quantum Theory of Measurement*. Lecture Notes in Physics. Berlin: Springer Verlag.

Cassettari, S. 1993. *Introduction to Integrated Geo-information Management*. London: Chapman & Hall.

Cattell, R. G. (ed). 1994. *The Object Database Standard: ODMG-93*. San Mateo, California: Morgan Kaufmann Publishers.

Chakravarthy, S. 1995. Eary active database efforts: A capsule summary. *IEEE Transactions on Knowledge and Data Engineering*, **7**(6), 1008–1010.

Chandy, K. M., & Rifkin, A. 1996. *Systematic composition of objects in distributed internet applications: Processes and sessions*. Tech. rept. 96-15. Caltech Computer Science.

Chrisman, N. R. 1983. The role of quality information in the long-term functioning of a geographic information system. *Cartographica*, **21**(2+3), 79–87.

Chrisman, N. R. 1991a. The error component in spatial data. *In:* (Maguire et al., 1991).

Chrisman, N. R. 1991b. Modeling error in overlaid categorical maps. *In:* (Goodchild & Gopal, 1991).

Chrisman, N. R. 1994. From a message in the NSDI-L mailing list.

Chrisman, N. R. 1995. Beyond Stevens: A revised approach to measurement for geographic information. *Pages 271–280 of: Proceedings of the AUTOCARTO 12 Conference.*

Clarke, D. G., & Clark, D. M. 1995. Lineage. *In:* (Guptill & Morrison, 1995).

Comer, D. E. 1991. *Internetworking with TCP/IP, Volume 1: Principles, protocols and architecture.* Second edn. Englewood Cliffs, New Jersey: Prentice-Hall.

Comer, D. E. 1995. *Internetworking with TCP/IP, Volume 1: Principles, protocols and architecture.* Third edn. Englewood Cliffs, New Jersey: Prentice-Hall.

Constantinescu, F. 1974. *Distributionen und ihre Anwendung in der Physik.* Stuttgart: Teubner.

Cook, S., & Daniels, J. 1994. *Designing Object Systems: object-oriented modelling with syntropy.* Englewood Cliffs, New Jersey: Prentice-Hall.

CORBA. 1992. *The Common Object Request Broker: Architecture and Specification.* Object Management Group, Framingham, Massachussetts. OMG Document Number 91.12.1.

CORBA2. 1995. *The Common Object Request Broker: Architecture and Specification.* Object Management Group, Framingham, Massachussetts. Version 2.0.

Couclelis, H. 1992. People manipulate objects (but cultivate fields): beyond the raster-vector debate in GIS. *Pages 65–77 of:* Frank, A. U., Campari, I., & Formentini, U. (eds), *Theories and Methods of Spatio-temporal Reasoning in Geographic Space.* Lecture notes in computer science, no. 639. Berlin: Springer Verlag.

Coulouris, G., Dollimore, J., & Kindberg, T. 1994. *Distributed Systems; concepts and design.* Second edn. Reading, Massachusetts: Addison-Wesley.

Daintith, J., & Nelson, R. D. (eds). 1991. *Dictionary of Mathematics.* London, United Kingdom: Penguin Books.

DCE/RPC. 1994. *DCE: Remote Procedure Call.* X/Open Company Limited, Reading, United Kingdom. X/Open Document Number C3091521.

De Lorenzi, M., & Wolf, A. 1993. A protocol for cooperative spatial information managers. *In: Workshop on Interoperability of Database Systems and Database Applications, Fribourg, Switzerland.*

Destouches, J. L. 1975. On basic problems about measurements and errors. *In: Proceedings of the symposium on Measurement Theory – Measurement Error Analysis.* IMEKO Secretariat.

Deutsch, C. V., & Journel, A. G. 1992. *GSLIB: Geostatistical Software Library and User's Guide*. Oxford: Oxford University Press.

DIGEST. 1991. *Digital Geographic Information Exchange Standard (DIGEST)*. Digital Geographic Information Working Group, Defense Mapping Agency.

Dijkstra, E. W. 1978. On the GREEN language submitted to the DoD. *ACM Sigplan Notices*, **13**(10).

Dimitrova, N. S., Markov, S. M., & Popova, E. D. 1992. Extended interval arithmetics: new results and applications. *In:* Atanassova, L., & Herzberger, J. (eds), *Computer Arithmetic and Enclosure Methods*. Amsterdam: Elsevier Science.

Djokic, D. 1996. Generic Data Exchange – Integrating Models and Data Providers. *In:* (NCGIA, 1996). Published on CD-ROM.

Doviak, R. J., & Zrnić, D. S. 1984. *Doppler Radar and Weather Observations*. Orlando, Florida: Academic Press.

Drummond, J. 1995. Positional accuracy. *In:* (Guptill & Morrison, 1995).

Dutton, G. 1996. Personal communication.

Egenhofer, M., & Herring, J. R. (eds). 1995. *Advances in Spatial Databases*. Lecture notes in computer science, no. 951. Berlin: Springer Verlag.

Ehlers, M., Greenlee, D., Smith, T., & Star, J. 1991. Integration of remote sensing and GIS: Data and data access. *Photogrammetric Engineering & Remote Sensing*, **57**(6), 669–675.

Englund, E. J. 1990. A variance of geostatisticians. *Mathematical Geology*, **22**(4), 417–455.

FGDC. 1994. *Content standard for digital geospatial metadata*. US Federal Geographic Data Committee, Reston, Virginia.

Finkelstein, L., & Carson, E. R. 1975. Errors due to incorrect model structure in measurements by parameter estimation. *In: Proceedings of the symposium on Measurement Theory – Measurement Error Analysis*. IMEKO Secretariat.

Firesmith, D. G., & Eykholt, E. M. 1995. *Dictionary of Object Technology*. New York: SIGS Books, Inc.

Fischbach, R. 1996. Kalter Kaffee. *iX Multiuser Multitasking Magazin*, **96**(10), 84–89.

Flanagan, D. 1996. *Java in a Nutshell*. Sebastopol, California: O'Reilly & Associates, Inc.

Flowerdew, R. 1991. Spatial data integration. *In:* (Maguire *et al.*, 1991).

Foresman, T., & McGwire, K. 1995. Data issues. *Pages 15–20 of:* Goodchild, M. F., Estes, J. E., Beard, K., Foresman, T., & Robinson, J. (eds), *Multiple Roles for GIS in US Global Change Research. Report of the First Specialist Meeting, Santa Barbara, CA, March 9–11*. Santa Barbara, CA: National Center for Geographic Information and Analysis. Technical report 95-10.

Frank, A.. U., & Kuhn, W. 1995. Specifying Open GIS with functional languages. *In:* (Egenhofer & Herring, 1995).

Franke, R. 1990. Approximation of scattered data for meteorological applications. *Pages 107–120 of:* Haussman, W., & Jetter, K. (eds), *Multivariate Approximation and Interpolation. Proceedings of an International Workshop at the University of Duisburg, August 14–18, 1989*. Basel: Birkhäuser.

French, J. C., & Hinterberger, H. (eds). 1994. *Seventh International Working Conference on Scientific and Statistical Database Management, September 28–30*. IEEE Computer Society Press.

Gal, A., & Etzion, O. 1995. Maintaining data-driven rules in databases. *Computer*, **28**(1), 28–38.

Gardels, K. 1996. *Coverage Type Hierarchies and Interfaces*. Tech. rept. OpenGIS Project Document 95-037R2. Open GIS Consortium, Wayland, Massachusetts.

Geörg, S., Hammer, R., Neaga, M., Ratz, D., & Shiriaev, D. 1993 (May). *PASCAL-XSC, a Pascal Extension for Scientific Computation and Numerical Data Processing*. Tech. rept. Institute for Applied Mathematics, University of Karlsruhe, Germany.

Gerasimov, V.A., Dobronets, B.S., & Shustrov, M. Y. 1991. Numerical operations of histogram arithmetic and their applications. *Automation and Remote Control*, **52**(2), 208–212.

Gilgen, H., & Steiger, D. 1992. The BSRN database. *In:* (Hinterberger & French, 1992).

Golder, P. A. 1990. Semantic modeling of bivariate statistical tests. *In:* (Michalewicz, 1990).

Goodchild, M. F. 1991. Issues of quality and uncertainty. *Pages 113–139 of: Advances in Cartography*. Amsterdam: Elsevier Science.

Goodchild, M. F. 1992. Geographical data modeling. *Computers and Geosciences*, **4**(18), 401–408.

Goodchild, M. F. 1995. Attribute accuracy. *In:* (Guptill & Morrison, 1995).

Goodchild, M.F., & Gopal, S. (eds). 1991. *Accuracy of Spatial Databases.* London, United Kingdom: Taylor & Francis.

Gosling, J., & McGilton, H. 1995. *The Java Language Environment.* Tech. rept. Sun Microsystems Computer Company, Mountain View, California.

Guptill, S. C. 1995. Temporal information. *In:* (Guptill & Morrison, 1995).

Guptill, S. C., & Morrison, J. L. (eds). 1995. *Elements of Spatial Data Quality.* Amsterdam: Elsevier Science.

Gust, P. J. 1990. The evolving role of user interface for inter-operability. *Pages 15–22 of:* Bron, C. A., Rasdorf, W. J., & Fulton, R. E. (eds), *Engineering Data Management: The technology for integration.* The American Society of Mechanical Engineers, New York, N.Y. Proceedings of the 1990 ASME international computers in engineering conference and exposition, August 5–9, Boston, Massachusetts.

Hanson, E. N. 1996. The design and implementation of the Ariel active database rule system. *IEEE Transactions on Knowledge and Data Engineering,* **8**(1), 157–172.

Härtig, M., & Dittrich, K. R. 1992 (Oct.). *An object-oriented integration framework for building heterogeneous database systems.* Tech. rept. 92.17. Institut für Informatik, Universität Zürich.

Heuvelink, G. B. M. 1993. *Error propagation in quantitative spatial modelling applications in Geographical Information Systems.* Doctoral Thesis, University of Utrecht.

Hiller, W., & Käse, R. H. 1983. *Objective analysis of hydrographic data sets from mesoscale surveys.* Berichte aus dem Institut für Meereskunde 116. Christian-Albrechts-Universität Kiel.

Hinterberger, H., & French, J. C. (eds). 1992. *Proceedings of the Sixth International Working Conference on Scientific and Statistical Database Management, June 9–12.* Departement Informatik, ETH Zürich, Zürich, Switzerland.

Hinterberger, H., Meier, K. A., & Gilgen, H. 1994. Spatial data reallocation based on multidimensional range queries. *In:* (French & Hinterberger, 1994).

Hinton, J. C. 1996. GIS and remote sensing integration for environmental applications. *International Journal of Geographical Information Systems,* **10**(7), 877–890.

Hofer, E. P., & Tibken, B. 1992. Simulation of Uncertain Discrete Time Systems with an Application to a Nonlinear Biomathematical Model. *Pages 111–121 of:* Kurzhanski, A. B., & Veliov, V. M. (eds), *Modeling Techniques for Uncertain Systems*. Basel: Birkhäuser.

Hofmann, D. 1975. Zur methodischen und mathematischen Behandlung von Messfehlern in der Labor-, Fertigungs- und Prozessmesstechnik. *In: Proceedings of the symposium on Measurement Theory – Measurement Error Analysis*. IMEKO Secretariat.

Hunter, G. J. 1992. Techniques for assessing the effect of processing errors in spatial databases. *Pages 24–33 of: Making Connections: URISA '92 Conference Proceedings*. Urban & Regional Information Systems Association.

IEEE. 1987. *Standard 754-1985 for Binary Floating-Point Arithmetic*. Institute of Electrical and Electronic Engineers, New York.

INTERLIS. 1990 (June). *INTERLIS: a Data Exchange Mechanism for Land-Information Systems*. Version 1.0e.

Isaaks, E.H., & Srivastava, M.R. 1989. *An Introduction to Applied Geostatistics*. Oxford: Oxford University Press.

Javasoft. 1996. *JavaBeans – Component APIs for Java.* http://www.javasoft.com/beans.

Jerrard, H. G., & McNeill, D. B. 1980. *A Dictionary of Scientific Units*. Fourth edn. London: Chapman & Hall.

Johannesson, P. 1993. *Schema integration, schema translation, and interoperability in federated information systems*. Doctoral Thesis, Department of Computer and Systems Sciences, Stockholm University.

Johnson, M. E. 1987. *Multivariate Statistical Simulation*. Wiley series in probability and mathematical statistics. Applied probability and statistics. New York: John Wiley and Sons.

Kainz, W. 1995. Logical consistency. *In:* (Guptill & Morrison, 1995).

Kalman, R. E. 1982. Identification from real data. *Pages 161–196 of:* Hazewinkel, M., & Kan, A. H. G. Rinnoy (eds), *Current Developments in the Interface: Economics, Econometrics, Mathematics*. Dordrecht: D. Reidel Publishing Company.

Kamke, D., & Krämer, K. 1977. *Physikalische Grundlagen der Messeinheiten: mit einem Anhang über Fehlerrechnung*. Stuttgart: Teubner.

Kapetanios, E. 1994. Anforderungen an die Datenbanktechnologie zur Verwaltung von Satellitendaten für die Fernerkundung atmosphärischer Parameter (Klimaforschung). *Pages 87–109 of:* Kremers, H. (ed), *Umweltdatenbanken.* Marburg: Metropolis-Verlag.

Kaucher, E., & Schumacher, G. n.d.. *Numerische Berechnung und Verifikation von Mittelwert-, Varianz und Konfidenzmengen für die Lösung statistisch gestörter Gleichungen.* unpublished.

Kemp, K. K. 1993 (May). *Environmental modelling with GIS: A strategy for dealing with spatial continuity.* Tech. rept. 93-3. National Center for Geographic Information and Analysis, Santa Barbara, California.

Klir, G. J. 1994. Measures of uncertainty in the Dempster–Shafer theory of evidence. *Chap. 2, pages 35–49 of:* Yager, R., Fedrizzi, M., & Kacprzyk, J. (eds), *Advances in the Dempster–Shafer Theory of Evidence.* New York: John Wiley and Sons.

Kolmogorov, A. N. 1933. *Grundbegriffe der Wahrscheinlichkeitsrechnung.* Berlin: Springer Verlag.

Krol, E. 1994. *The Whole Internet User's Guide & Catalog.* Second edn. Sebastopol, California: O'Reilly & Associates, Inc.

Kyburg, H. E. 1992. Measuring Errors of Measurement. *Chap. 5, pages 75–91 of:* Savage, C. W., & Ehrlich, P. (eds), *Philosophical and Foundational Issues In Measurement Theory.* Hillsdale, New Jersey: Lawrence Erlbaum Associates, Inc.

Lam, N. S. 1983. Spatial interpolation methods: A review. *American Cartographer*, **10**, 129–149.

Lanter, D. L. 1990. *The problem and a solution to lineage in GIS.* Tech. rept. 90-6. National Center for Geographic Information and Analysis, Santa Barbara, California.

Lauer, D. T., Estes, J. E., Jensen, J. R., & Greenlee, D. D. 1991. Institutional issues affecting the integration and use of remotely sensed data and geographic information systems. *Photogrammetric Engineering & Remote Sensing*, **57**(6), 647–654.

Laurini, R. 1996. Spatial multidatabase indexing and topological continuity of fragmented geographic objects. *In:* (Yétongnon & Hariri, 1996).

Leaning, M. S., Carson, E. R., Cobelli, C., & Finkelstein, L. 1984 (May). Validity of measurements obtained by model identification and parameter estimation. *Pages 83–86 of: Proceedings of the symposium on measurement and estimation, Bressanone, Italy.* IMEKO Technical Commitee on Measurement Theory.

Leclercq, E., Benslimane, D., & Yétongnon, K. 1996. A distributed object architecture for interoperable GIS. *In:* (Yétongnon & Hariri, 1996).

Leffler, S., McKusick, M., Karels, M., & Quartermain, J. 1989. *The Design and Implementation of the 4.3 BSD UNIX Operating System.* Reading, Massachusetts: Addison-Wesley.

Lenz, H. J. 1994. The conceptual schema and external schemata of metadatabases. *In:* (French & Hinterberger, 1994).

Litwin, W., Mark, L., & Roussopoulos, N. 1990. Interoperability of multiple autonomous databases. *ACM Computing Surveys*, **22**(3), 267–293.

Lodwick, W. A. 1991. Developing confidence limits on errors of suitability analyses in geographical information systems. *In:* (Goodchild & Gopal, 1991).

Lucent. 1996. *Lucent Technologies' Inferno.* http://www.lucent.com/inferno.

Lunetta, R. S., Congalton, R. G., Fenstermaker, L. K., Jensen, J. R., McGwire, K. C., & Tinney, L. R. 1991. Remote sensing and geographic system data integration: Error sources and reserach issues. *Photogrammetric Engineering & Remote Sensing*, **57**(6), 677–687.

MacEachren, A. M. 1992. Visualizing uncertain information. *Cartographic Perspectives*, 10–19.

Mackaness, W., & Beard, K. 1993. Visualization of interpolation accuracy. *Pages 228–237 of: Proceedings of the AUTOCARTO 11 Conference.*

Maffini, G., Arno, M., & Bitterlich, W. 1991. Observations and comments on the generation and treatment of error in digital GIS. *In:* (Goodchild & Gopal, 1991).

Maguire, D. J. 1991. An overview and definition of GIS. *In:* (Maguire *et al.*, 1991).

Maguire, D. J., Goodchild, M. F., & Rhind, D. W. (eds). 1991. *Geographical Information Systems: Princinples and applications.* London, United Kingdom: Longman.

Mason, D. C., O'Conaill, M. A., & Bell, S. B. M. 1994. Handling four-dimensional geo-referenced data in environmental GIS. *International Journal of Geographical Information Systems*, **8**(2), 191–215.

Mauch, S., Keller, M., Heldstab, J., Rotach, M., & Ramseier, H. 1992. *Fehlerrechnung und Sensitivitätsanalyse für Fragen der Luftreinhaltung.* SVI-Forschungsauftrag 46/90. Vereinigung Schweizerischer Verkehrsingenieure SVI.

Maxwell, T., & Costanza, R. 1996. Facilitating High Performance, Collaborative Spatial Modeling. *In:* (NCGIA, 1996). Published on CD-ROM.

Mayer, G. 1989. Grundbegriffe der Intervallrechnung. *In:* Kulisch, U. (ed), *Wissenschaftliches Rechnen mit Ergebnisverifikation*. Berlin: Akademieverlag.

Mermin, N. David. 1985. Is the moon there when nobody looks? Reality and the quantum theory. *Physics Today*, April, 38–47.

Meyer, B. 1992. Applying "design by contract". *Computer*, **25**(10), 40–51.

Michalewicz, Z. (ed). 1990. *Statistical and scientific database management: Fifth International Conference on Statistical and Scientific Databases, Charlotte, North Carolina, April 3-5*. Lecture notes in computer science, no. 420. Berlin: Springer Verlag.

Moellering, H. 1991. *Spatial Data Transfer Standards: Current international status*. Amsterdam: Elsevier Science.

Moore, R. 1966. *Interval Analysis*. Englewood Cliffs, New Jersey: Prentice-Hall.

Moore, R. 1992. Parameter sets for bounded-error data. *Mathematics and Computers in Simulation*, **34**, 113–119.

Morrison, J. 1988. The proposed standard for digital cartographic data. *The American Cartographer*, **15**(1), 129–135.

Nagel, E., & Newman, J. R. 1992. *Der Gödelsche Beweis*. München: R. Oldenbourg Verlag.

Narens, L. 1985. *Abstract measurement theory*. Cambridge, Massachusetts: The MIT Press.

NCGIA. 1996. *Proceedings of the Third International Conference/Workshop on Integrating GIS and Environmental Modeling, Santa Fe, NM, January 21-26*. National Center for Geographic Information and Analysis, Santa Barbara, CA. Published on CD-ROM.

Neumaier, A. 1993. The wrapping effect, ellopsoid arithmetic, stability and confidence regions. *Pages 175-190 of:* Albrecht, R., Alefeld, G., & Stettler, H.J. (eds), *Validation Numerics: Theory and Applications*. Berlin: Springer Verlag.

ODBC. 1992. *Microsoft Open Database Connectivity Software Development Kit: Programmer's Reference*. Microsoft Corporation, Redmond, Washington.

OGIS. 1996. *The OpenGIS Abstract Specification: An Object Model for Interoperable Geoprocessing, Revision 1.* Open GIS Consortium, Wayland, Massachusetts.

O'Rourke, J. 1994. *Computatinal Geometry in C.* New York: Cambridge University Press.

Oswald, R. 1993. The information utility. *Dr. Dobb's Journal,* **18**(13), 18–30.

Otte, R., Partick, P., & Mark, R. 1996. *Understanding CORBA.* Englewood Cliffs, New Jersey: Prentice-Hall.

Özsoyoğlu, G., & Snodgrass, R. T. 1995. Temporal and real-time databases: A survey. *IEEE Transactions on Knowledge and Data Engineering,* **7**(4), 513–532.

Parent, C., Spaccapietra, S., & Devogele, T. 1996. Conflicts in spatial database integration. *In:* (Yétongnon & Hariri, 1996).

Pfaltz, J. L., & French, J. C. 1990. Implementing subscripted identifiers in scientific databases. *In:* (Michalewicz, 1990).

Pilz, J. 1993. Some thoughts on the present position in Bayesian statistics. *Pages 70–82 of:* Bandemer, H. (ed), *Modelling Uncertain Data.* Mathematical research, no. 68. Berlin: Akademieverlag.

Pinson, L. J., & Wiener, R. S. 1991. *Objective-C: Object-oriented programming techniques.* Reading, Massachusetts: Addison-Wesley.

Piotrowski, J. 1992. *Theory of Physical and Technical Measurement.* Amsterdam: Elsevier Science.

Polyak, B., Scherbakov, P., & Shmulyian, S. 1992. Circular arithemtic and its applications in robustness analysis. *Pages 229–243 of:* Kurzhanski, A. B., & Veliov, V. M. (eds), *Modeling Techniques for Uncertain Systems.* Basel: Birkhäuser.

Qiu, K., Hachem, N., Ward, M. O., & Gennert, M. A. 1992. Providing temporal support in data base management systems for global change research. *In:* (Hinterberger & French, 1992).

Rhind, D.W., Green, N.P.A., Mounsey, H.M., & Wiggins, J.C. 1984. The integration of geographical data. *Pages 237–253 of: Proceedings of Austra Carto, Perth.* Oxford: Oxford University Press, for Australian Institute of Cartographers.

Ridley, B. K. 1995. *Time, Space and Things.* Third edn. Cambridge: Cambridge University press.

Salgé, F. 1995. Semantic accuracy. *In:* (Guptill & Morrison, 1995).

Schmerling, S., & Bandemer, H. 1985. Methods to estimate parameters in explicit functional relationships. *Pages 69–89 of:* Bandemer, H. (ed), *Problems of Evaluation of Functional Relationships from Random Noise or Fuzzy Data.* Freiberger Forschungshefte, no. D 170. Dt. Verlag für Grundstoffindustrie.

Schmerling, S., & Bandemer, H. 1990. Estimation of functional connections from fuzzy observations. *Pages 27–42 of:* Bandemer, H. (ed), *Some Applications of Fuzzy Set Theory in Data Analysis II.* Freiberger Forschungshefte, no. D 197. Dt. Verlag für Grundstoffindustrie.

Schneider, B. 1995. Adaptive interpolation of digital terrain models. *Pages 2206–2210 of: Proceedings of the 17th International Cartographic Conference.* International Cartographic Association, Barcelona (Spain).

Schrefl, M. 1988. *Object-oriented database integration.* Dissertation, Technische Universität Wien.

Schulin, R., Flühler, H., Selim, H. M., Sevruk, B., & Wierenga, P.J. 1992. Mathematical Expectations and Moments. *Chap. Annex III of: Snow cover measurement and areal assessment of precipitation and soil moisture.* Operational Hydrology Report, no. 35. Secretariat of the World Meteorological Organization.

Shafer, G. 1976. *A Mathematical Theory of Evidence.* Princeton, New Jersey: Princeton University Press.

Shannon, C. 1948. A mathematical theory of communication. *Bell Systems Tech. J.*, **27**, 379–423.

Shaw, K., Cobb, M., Chung, M., & Arctur, D. 1996. Managing the US Navy's first OO digital mapping project. *Computer*, **29**(9), 69–74.

Shepherd, I. D. H. 1991. Information integration and GIS. *In:* (Maguire *et al.*, 1991).

Sheth, A. P., & Larson, J. A. 1990. Federated database systems for managing distributed, heterogeneous, and autonomous databases. *ACM Computing Surveys*, **22**(3), 183–236.

Smith, F. J., & Krishnamurthy, M. V. 1992. Integration of scientific data and formulae in an object-oriented system. *In:* (Hinterberger & French, 1992).

Smith, T. R., Su, J., Abbadi, A. El, Agrawal, D., Alonso, G., & Saran, A. 1995. Computational modeling systems. *Information Systems*, **20**(2), 127–153.

Sorokin, A., & Merzlyakova, I. 1996. *List of GIS standards and formats.* http://www.ru/gisa/english/cssitr/format/index.htm.

Stafford, P., & Powell, J. 1995. COM: A model problem solver. *Microsoft Developer Network News*, **4**(2), 1, 12.

Stanek, H., & Frank, A. U. 1993. GIS based decision making must consider data quality. *Pages 685–691 of: Proceedings of the EGIS'93 Conference*, vol. 1.

Star, J. L., Estes, J. E., & Davis, F. 1991. Improved integration of remote sensing and geographic information systems: A background to NCGIA initiative 12. *Photogrammetric Engineering & Remote Sensing*, **57**(6), 643–645.

Stephan, E.-M., Včkovski, A., & Bucher, F. 1993. Virtual Data Set: an approach for the integration of incompatible data. *Pages 93–102 of: Proceedings of the AUTOCARTO 11 Conference.*

Stevens, W. T. 1990. *UNIX Network Programming.* Englewood Cliffs, New Jersey: Prentice-Hall.

Stonebraker, M. 1994. Sequoia 2000 – A reflection on the first three years. *In:* (French & Hinterberger, 1994).

Stumptner, M. 1990. *Redundancy and information content of data relations with different kinds of null values.* Dissertation, Technische Universität Wien.

Teitelmann, W. 1984. A tour through Cedar. *IEEE Software*, **1**(2), 44–73.

Thiébaux, H.J., & Pedder, M.A. 1987. *Spatial Objective Analysis with Applications in Atmospheric Science.* London: Academic Press.

Thieme, C., & Siebes, A. 1993 (Mar.). *Schema integration in object-oriented databases.* Tech. rept. CS-R9320. CWI, Stichting Mathematisch Centrum, Amsterdam, Netherlands.

Tobler, W. R. 1979. Smooth pycnophylactic interpolation for geographical regions. *Journal of the American Statistical Association*, **74**, 519–530.

Valdés, R. 1994. Introducing interoperable objects. *Dr. Dobb's Journal*, **19**(16), 4–6.

van Delft, A. 1996. Dimensioned numbers in Java. *Java Developer's Journal*, **1**(1), 20–22.

van den Berg, G., & de Feber, E. 1992. Definition and use of meta-data in statistical data processing. *In:* (Hinterberger & French, 1992).

van Hoff, A., Shaio, S., & Starbuck, O. 1996. *Hooked on Java.* Reading, Massachusetts: Addison-Wesley.

Veregin, H. 1991. Error modelling for the map overlay operation. *In:* (Goodchild & Gopal, 1991).

Voisard, A., & Schweppe, H. 1994. A multilayer approach to the open GIS design problem. *Pages 23–29 of: Proceedings of the Second ACM GIS Workshop.*

Včkovski, A. 1995. Representation of Continuous Fields. *Pages 127–136 of: Proceedings of the AUTOCARTO 12 Conference.*

Včkovski, A. 1996. Java as a software system for distributed and interoperable geoprocessing. *In:* (Yétongnon & Hariri, 1996).

Včkovski, A., & Bucher, F. 1996. Virtual Data Sets – smart data for environmental applications. *In:* (NCGIA, 1996). Published on CD-ROM.

Wesseling, C. G., van Deursen, W. P. A., & Burrough, P. A. 1996. A spatial modelling language that unifies dynamic environmental models and GIS. *In:* (NCGIA, 1996). Published on CD-ROM.

Wheeler, D. J. 1993. Commentary: Linking environmental models with geographic information systems for global change research. *Photogrammetric Engineering & Remote Sensing*, **59**(10), 1497–1501.

Wiethoff, A. 1992 (Oct.). *C-XSC, a C++ class library for extended scientific computing.* Tech. rept. Institute for Applied Mathematics, University of Karlsruhe, Germany.

Wirth, N. 1996. Recollections about the development of Pascal. *In:* Bergin, T. J., & Gibson, R. G. (eds), *History of Programming Languages II.* New York: ACM Press.

Wirth, N., & Gutknecht, J. 1992. *Project Oberon: the design of an operating system and compiler.* New York: ACM Press.

Wolfram, S. 1988. *Mathematica, a system for doing mathematics by computers.* Reading, Massachusetts: Addison-Wesley.

Worboys, M. F. 1992. A model for spatio-temporal information. *Pages 602–611 of:* Bresnahan, P., Corwin, F., & Cowan, D. (eds), *Proceedings of the 5th International Symposium on Spatial Data Handling, Charleston, SC, 3–7 August.* International Geographical Union, Commission on GIS.

Worboys, M. F. 1994. Object-oriented approaches to geo-referenced information. *International Journal of Geographical Information Systems*, **8**(4), 385–399.

Worboys, M. F., & Deen, S. M. 1991. Semantic heterogeneity in distributed geographical databases. *SIGMOD Record*, **20**(4).

Yétongnon, K., & Hariri, S. (eds). 1996. *Proceedings of the Ninth International Conference on Parallel and Distributed Computing Systems, Dijon, France, September 25–27*. The International Society for Computers and Their Applications, Raleigh, North Carolina.

Zadeh, L. A. 1965. Fuzzy Sets. *Information and Control*, **8**, 338–353.

Index

accessor, 144, 150, 194
active database, 44
agent, 30
aggregation, 87
autocorrelation, 32, 88

cell grid, 77
client/server, 148
complexity, 40, 147, 159
content negotiation, 135
contours, 23, 82
CORBA, 148–154, 203

data definition language, 40, 150
data directory, 30
data exchange, 29
data integration, 1, 20, 201
data migration, 10
data model, 23, 74–86
data quality, 22, 26–28, 32
database integration, 14
dimension, 112, 183, 192
distributed computing, 170, 174, 203, 207
domain, 65, 89

environmental data, 31
error propagation, 27, 97, 106–110
executable content, 58, 121, 207
EXPRESS, 40, 150

expressiveness, 13
extrapolation, 26, 87

feature, 48
FGDC, 29
field view, 23, 48
fields, 32, 65–90, 143, 193–199
formal system, 45, 91
fuzzy sets, 103

generalization, 30

histogram, 103, 112, 188
HTML, 134
HTTP, 134, 155

IDL, 150
IEEE 754, 93
information community, 38, 47, 202
interdisciplinarity, 32
interface, 47, 145, 150, 162, 181, 207
INTERLIS, 40
Internet, 128–141
interoperability, 9, 35, 164, 181, 201
  case study, 119–141
interpolation, 26, 87
interprocess communication, 129, 170, 174
interval arithmetics, 100, 186

Java, 157–180

lattice, 80
legacy system, 31

market forces, 39
mathematical model, 91
measurement model, 67
metadata, 28, 32, 125
migration, 153
MIME, 135, 136, 155, 158
Monte Carlo methods, 109
multidatabase systems, 14
multiple representations, 58, 137

null value, 182, 195

object view, 23, 48
object-orientation, 41, 149, 157, 159, 202
objective analysis, 86
OGIS, 38, 46, 193, 204, 205
OLE/COM, 144–147, 203
open geodata model, 47

pixel, 70, 85
point grids, 80
polygon overlay, 25
polyhedral tessellation, 77, 85
POSTSCRIPT, 119–128
printer driver, 55
probability theory, 98, 186

quality assurance, 29

random field, 70
range, 65, 89
remote sensing, 21

scalability, 130
security, 177
selection function, 69
semantic accuracy, 45
semantic heterogeneity, 18, 22, 37, 202
simplicial complexes, 80

spatial join, 25, 32
standardization, 10, 138
support, 65
syntactical heterogeneity, 36
system integration, 30

uncertainty, 27, 91–115, 182
unit, 112, 183, 192
URL, 135, 155, 158

Virtual Data Set, 35, 54–60, 143, 174, 180–199, 207

World-Wide Web, 134–138, 157